Canyons of the Southwest

Canyons of the Southwest

A Tour of the Great Canyon Country from Colorado to Northern Mexico

PHOTOGRAPHS AND TEXT BY
John Annerino

THE UNIVERSITY OF ARIZONA PRESS · TUCSON

Author and photojournalist John Annerino was born on the edge of the desert and cut his teeth in the heart of the American West. He has been working in the frontier of Old Mexico and the American West for more than two decades, documenting its natural beauty, indigenous people, and political upheaval. A contract photographer with the Liaison International picture agency in New York and Paris, Annerino's credits include *Life, Time, Newsweek*, the *New York Times, Scientific American*, and many other prestigious publications worldwide. He is the author of thirteen books, including seven photo essays on this mythic region and the spirited individualists and Native people who have lived here since the time of legend. Annerino's recent works include *Apache: The Sacred Path to Womanhood, Roughstock: The Toughest Events in Rodeo, Running Wild,* and *Dead in Their Tracks: Crossing America's Desert Borderlands.*

We gratefully acknowledge the following for permission to quote from copyrighted material: Robert B. McCoy for the excerpt from *Unknown Mexico,* published in 1973 by The Rio Grande Press, Inc., Glorieta, New Mexico; United States Department of the Interior for the excerpt from *The Big Bend—A History of the Last Texas Frontier,* a National Park Service handbook; Time-Life Books Inc. for the excerpt from *The American Wilderness: Baja California,* by William Weber Johnson and the editors of Time-Life Books, Copyright ©1972 by Time-Life Books; and Random House, Inc. for the excerpt from *The Inverted Mountains: Canyons of the West,* by Roderick Peattie, Copyright © 1948 by Roderick Peattie.

First University of Arizona Press paperbound edition
The University of Arizona Press
Photographs and text copyright © 1993 by John Annerino
All rights reserved
♾ This book is printed on acid-free, archival-quality paper.
Manufactured in the United States of America

05 04 03 02 01 00 6 5 4 3 2 1

Library of Congress Cataloging-in-Publication Data
Annerino, John.
Canyons of the Southwest : a tour of the great canyon country from Colorado to
northern Mexico / photographs and text by John Annerino.— 1st University of
Arizona Press paperbound ed.
p. cm.
Originally published: San Francisco : Sierra Club Books, c1993.
Includes bibliographical references.
ISBN 0-8165-2092-5 (pbk. : alk. paper)
1. Canyons—Southwest, New. 2. Canyons—Mexico. I. Title.
GB566.S68 A56 2000
551.4'42—dc21   00-041785

British Library Cataloguing-in-Publication Data
A catalogue record for this book is available from the British Library.

Contents

Acknowledgments

Photographing a project of this scope would have been impossible without the many kind people who helped me along the way.

In the frontier of Chihuahua, I would like to thank Margarita Quintero de Gonzalez, Alfonso Saenz-Hernandez, Jesus Israel Gonzalez, Nacho Palma, Tony Jones, Andri Rauch, Shaun Cannon, Cristina Moreno, Elva Moreno-Chaparro and Nicolas Chaparro-Moreno, Ulrike Wulf, Elva Soto-Gill, and Domingo Uriarte Parra.

In the frontier of Coahuila and the Texas Big Bend country, I would like to thank José and Ophelia Falcon, Lucia Rede-Madrid, Rafael Carrasco, Gus Sanchez, Miguel Ureste, Candelario Valdez, Maria Sanchez, Tano Padilla, Teresa Padilla-Ureste, Ramona Padilla-Rodriguez, Toro and Chero, Juan Valdez, Victor Valdez, Silvia Ureste, Jacqueline Falcon-Jimenez, Ricardo Moran-Espinoza, Miguel Ureste-Luna, Curt Gutherie, Gary Joyce, Johnny Holland, Ventura Orozco, and Kiko Garcia.

I would like to thank Santiago, Ernesto Molino, Neil Carmony, Dave Brown, and Roseanne and Jonathan Hanson for our adventures on Isla Tiburón.

And I would also like to thank Jason Lohman, Eric Lohman, Ret. Major Bruce Lohman, Aimee Madsen, Wayne Von Voorhees, John Dell, Nancy Thompson, Richard Nebeker, Bob Farrell, Mike Thomas, Dave Ganci, Tony Mangine and family, Chris May, Chris Keith, Galen Snell, George Bain, Louise Teal, Martha Clark, Suzanne Jordon, and Rob Elliott for also sharing some of my more memorable canyon adventures.

I appreciate the encouragement of my peers like Melvin L. Scott, Richard Vonier, Elvira Mendoza, and Charlotte Thompson and am grateful for the help I received from the kind staff at the University of Arizona's Main, Science, and Special Collections libraries and for the help I received from the research librarians at the Arizona Historical Society in Tucson.

I am grateful for the fine work of designer Paula Schlosser and mapmaker Hilda Chen, as well as that of University of Arizona Press director Christine Szuter and editorial assistant Julianne Blackwell.

And I am indebted to native speaker, friend, and language professor Lucinda Bush-Garrett for making certain my use of Spanish grammar and punctuation was correct.

And last, but not least, I would like to thank the native people of the Navajo, Ute, Southern Paiute, Yavapai, Apache, Hopi, Havasupai, Walapai, Kiliwa, Seri, Comanche, and Tarahumara nations for opening my eyes to their life—and spiritways. Thank you.

Introduction

Shouts are lost in the immensity of space and chasm of the
canyons. . . . Here is an occasion . . . for the beating of great
drums, and, if for dancing, a dance in which one leaps into the
air. There is nothing like the Canyon Country in all the world.

RODERICK PEATTIE
The Inverted Mountains: Canyons of the West (1948)

There *isn't* anything like it in all the world. The Colca Canyon in Peru is
thought to be the deepest at 14,339 feet, and the Tiger Leaping Gorge in
China is said to be wilder, but nowhere else on earth is there a canyon region
that compares to North America's Southwest. Here are found the narrowest
canyons on earth, the largest, the most spectacular. But superlatives alone
cannot describe them. For this is the mythical Southwest, a vast legendary
sweep of land that encompasses Utah, Colorado, Arizona, New Mexico,
Texas, and, south of the U.S.–Mexico border, the states of Baja California
Norte, Sonora, Chihuahua, Sinaloa, and Coahuila. But, unlike the region's
great mountain ranges, it is a storied land not readily seen if you are standing
on flat ground. You must go to the edge of a wind-dusted plateau to see the
magnificent world that exists below its horizon line; or you must climb to
the summit of a remote sierra to see the tangled labyrinth of canyons that fan
out from its crest. In either case, you needn't leave your chair now, because

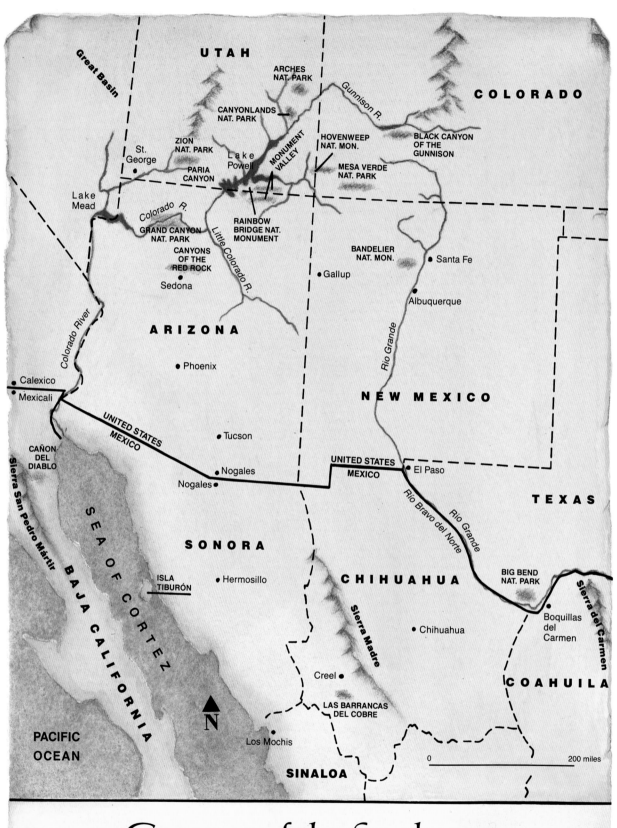

UTAH

COLORADO

Great Basin

ARCHES NAT. PARK

Gunnison R.

CANYONLANDS NAT. PARK

ZION NAT. PARK

St. George

Lake Powell

MONUMENT VALLEY

HOVENWEEP NAT. MON.

BLACK CANYON OF THE GUNNISON

PARIA CANYON

MESA VERDE NAT. PARK

Lake Mead

Colorado R.

GRAND CANYON NAT. PARK

Little Colorado R.

RAINBOW BRIDGE NAT. MONUMENT

BANDELIER NAT. MON.

Santa Fe

CANYONS OF THE RED ROCK

Gallup

Albuquerque

Sedona

ARIZONA

Colorado River

Phoenix

Rio Grande

NEW MEXICO

Calexico

Mexicali

UNITED STATES
MEXICO

Tucson

CAÑON DEL DIABLO

UNITED STATES
MEXICO

El Paso

Nogales

TEXAS

Sierra San Pedro Mártir

Nogales

Rio Bravo del Norte

Rio Grande

SEA OF CORTEZ

SONORA

CHIHUAHUA

BIG BEND NAT. PARK

ISLA TIBURÓN

Hermosillo

Sierra del Carmen

BAJA CALIFORNIA

Sierra Madre

Chihuahua

Boquillas del Carmen

N

Creel

COAHUILA

PACIFIC OCEAN

LAS BARRANCAS DEL COBRE

Los Mochis

0   200 miles

SINALOA

Canyons of the Southwest

Canyons of the Southwest will take you on a pictorial, adventure-filled journey into canyons and chasms at once sacred and sublime.

Encompassing 130,000 square miles of Colorado, Utah, New Mexico, and Arizona, the 6,000-foot-high Colorado Plateau has the most spectacular canyons in the United States. Its principal cutting agent, the Colorado River, drains an estimated 244,000 square miles; through a combination of wind, erosion, and the ancient upheaval of great land masses, it has sculpted the slickrock canyon lands of Utah, the deep, haunting fissures of Buckskin Gulch and Paria Canyon, and the greatest of them all, the Grand Canyon of the Colorado River.

In *The Inverted Mountains,* Roderick Peattie described the Grand Canyon as "the main range—the Mount Everest of the Inverted Mountains." He was right about that. Because even its awesome tributary canyons like the Little Colorado River Gorge, Kanab Creek, and the double canyons of Cataract and Havasu overshadow most of the region's other canyons in both stature and sheer grandeur.

Take Colorado's Black Canyon of the Gunnison, on the eastern periphery of the Colorado Plateau, for one; nearly 3,000 feet deep and 12 miles long, the heart of Black Canyon would easily fit into the lower one-third of any of the Grand Canyon's three largest tributaries. But it's been included in what boiled down to a difficult selection of canyons because the Black Canyon of the Gunnison—not the great Rocky Mountains that loom above it—boasts the tallest, most imposing wall in Colorado: the 2,300-foot-high Painted Wall.

On the western edge of the Colorado Plateau, on the other hand, you will trek down the length of Zion Narrows. At sixteen miles long, it is little more than a footnote compared to the Grand Canyon. It was included not only for its scenic beauty but also because it was sacred to both the Southern Paiute, who once revered it, and to Mormon pioneers who later settled the area. That's also why Bandelier National Monument in New Mexico, Mesa Verde National Park in Colorado, and Hovenweap National Monument in Utah have been included—because of their sanctity to native peoples, not because their small canyons can compare in size to either the Little Colorado River Gorge, Kanab Creek, or Cataract and Havasu canyons. Even with some 2,500 ruins catalogued in the depths of the Grand Canyon alone, none compare to those of Hovenweap, Mesa Verde, Bandelier, and elsewhere; because nowhere else on the Colorado Plateau is there such a spectacular concentration of sacred ruins once inhabited by the Anasazi, "the Old Ones," one of the earliest peoples who inhabited the canyon lands of the Colorado Plateau, as those still clinging to the walls of these small outlier canyons.

No exploration of the Colorado Plateau would be complete, however, without including the canyons that incise its 200-mile-long southern escarpment, called the Mogollon Rim. Boulder-choked canyons such as

Sycamore and the watery courses of West Fork of Oak Creek and Wet Beaver Creek, to name a few, not only bridge the threshold between the Colorado Plateau and the Basin and Range physiographic provinces, but they were also home to native people who lived peacefully within their canyon environments much as the Havasupai still do in the western Grand Canyon.

To say canyons solely exist in the Colorado Plateau region, as many books tell us, is to believe the Southwest is a region that solely exists north of the U.S.–Mexico border. South of this demarcation line, however, there are canyons as wild, rugged, and beautiful as anything found north of it. Once the haunt of native Kiliwa Indians and Mexican miners, Cañón del Diablo leads to the summit of the highest mountain in Baja—the 10,154-foot summit of Picacho del Diablo, Peak of the Devil. Venture southeast from the nearby gulf port of San Felipe across the Sea of Cortez in flimsy reed *balsas* to Isla Tiburón, as the Seri Indians once did, and you will explore an island canyon once visited by Thor Heyerdahl, who journeyed there to investigate the use of reed boats before embarking on his famous Ra Expeditions.

Five suns east of the Seri's ancestral land, however, the Canyons of the Big Bend must also be reckoned with. The Río Grande has carved a canyon system and nurtured a bioregion that has always linked the United States and Mexico and, because of this uniqueness, is currently being studied as an international park. Coahuila's Sierra del Carmen links Mexico's Sierra Madre Oriental with the Rocky Mountains and Cañón del Boquillas is one of the sole breaches through this great transcontinental range.

No exploration of the canyons of the Southwest would be complete, however, without including the Barrancas del Cobre region of Chihuahua and Sinaloa. This is the one canyon system that is most often described as being deeper and larger than the Grand Canyon of the Colorado River. Whether the Sierra Madre's magnificent rivers have created any *barrancas*, canyons, deeper than the Grand Canyon is doubtful. Of the region's five deepest canyons, the Barranca del Urique has been called the deepest at 6,136 feet, while the Barranca de Sinafarosa has been measured at 6,002 feet deep. Bottoming out close behind is the 5,904-foot-deep Barranca del Batopilas; the 5,770-foot-deep Barranca del Cobre, "Copper Canyon," from which the region takes its popular name; and the 5,313-foot-deep Barranca del Guaynopa. At what point these measurements were taken, however, is not clearly known and relying solely on the 1:500,000 topographical maps produced by Mexico's Departamento Cartográfico, these canyons may actually be deeper—or even shallower. But for the sake of comparison, if these figures are dead on, the Grand Canyon is still deeper. From the 2,369-foot level at Granite Rapids at Mile 95.5 on the Colorado River to the 9,089-foot contour on Crystal Ridge atop the North Rim, the Granite Gorge section of the Grand Canyon is 6,720 feet deep; while the Marble Canyon section of the Grand Canyon is 6,463 feet deep, measured from the 2,763-foot level of Kwagunt Rapid at River Mile 56 to the North Rim near the head of South Canyon. Even so, only Idaho's 7,900-foot-deep Hell's Canyon is undisputedly deeper than all of North America's greatest canyons.

In the same breath Las Barrancas del Cobre has been touted as being deeper than the Grand Canyon, it's also been said to be "three to four times larger." But that's comparing apples to oranges. Any of Las Barrancas del Cobre's immense gorges would fit lengthwise into the 277-mile-long Grand Canyon of the Colorado River. Widthwise, their "rims" are not as easily as defined as the Grand Canyon's. So how do you compare the mountain canyon system of the Sierra Madre to the plateau canyon system of the Colorado Plateau, if in fact you can? As a region comprising the existing lands of the Tarahumara Indians, Las Barrancas del Cobre— including the occupied highlands of its Cordilleran Plateau Province—has been estimated to be 25,000 square miles; taken solely by itself, however, the official national park boundaries of the Grand Canyon only comprise some 2,000 square miles. But if you looked at the Grand Canyon as being just one of several stupendous gorges that comprise the canyons of the Colorado River system (including Desolation, Gray, Stillwater, and Cataract canyons, among many others), much the way the canyons of Las Barrancas del Cobre form their own system, then you have to consider how much of the 130,000-square-mile region of the Colorado Plateau you want to include in the debate.

Be that as it may, if in fact the canyons of the Colorado River are deeper, and larger, and geologically more colorful than the Barrancas del Cobre region, they pale in comparison to the number of indigenous people still living in the Barrancas del Cobre region, much as the Tarahumara did 2,000 years ago. So if the Grand Canyon is the one great canyon against which all others are measured, the Barrancas del Cobre is still the only window we now have to see how native peoples once inhabited all the other great canyons of the Southwest. For this reason we've included in this pictorial odyssey the great Tarahumara people; they are living proof that, north of the U.S.–Mexico border, native people once lived in harmony with their ancestral canyon lands. And that is reason enough to dance.

To paraphrase Peattie: here in the canyons of the Southwest, "Shouts are still lost in the immensity of space and chasm of the canyons." But it's in the *barrancas* of the Tarahumara that you can still hear the beating of their great drums, that you can still see the sacred dancing in which they leap into the air. And fortunately, still, there is nothing like this canyon country in all the world.

Come see for yourself.

Canyon Lands of the High Plateau
Utah and Arizona

Stand alone on the summit of some sacred mountain and look
far out over the weird rocks and canyons and sagebrush flats.
It would not be surprising to see a Yei at the foot of yonder
rainbow, preparing to travel on it with some hero of legend.

EDITH L. WATSON
Navajo Sacred Places (1964)

It would not be surprising at all. Because all the rocks have names. All the canyons have names. And all the stone rainbows have names. Especially Nonnezoshi; it is "Rainbow Rock-Span," and it is sacred. Most people don't know that, and most don't know the land around it is sacred: the hoodoo spires; the slickrock dungeons; the deadrock deserts. Yet in the setting sun it appears little more than a vision of naked, flame-red stone: hundreds of miles of strange towers, serpentine mazes, winged arches, horizonless badlands, riprap canyons that pry open a tortured and mythical ground, and the lofty ranges of the Henrys, the La Sals, and the Abajos that soar above a world hidden beneath the horizon line. But there *is* more; there always is. The clues lie in the names: Ear of the Wind, Sun's Eye, and Rain God Mesa. They tell a story that most can never know, because it is the sacred knowledge of native peoples who have always known these mystical canyon lands as home.

1

Beyond the shadow of this twilight reverie, there is more—30,000 square miles more—because the land beyond Nonnezoshi forms the canyon lands of the High Plateau, an unusually desolate, abandoned tract of the 6,000-foot-high Colorado Plateau. It is rimmed on the north by a fossilized wave of rock that now takes the name of Utah's Book Cliffs; on the west, it's haunted by the Box-Death Hollow ground of the Aquarius Plateau; on the south, it drifts into the mind-warping mirage Spaniards first called the *Desierto Pintado,* "Painted Desert." But it's in the direction of the east wind that this desert ground was truly formed—by the Ice Age waters born on the spine of an ancient continent.

First among them is the San Juan River. It trickles down out of the 13,232-foot La Plata Mountains of Colorado and for 360 miles snakes its way across northwest New Mexico and southeast Utah, falling 9,000 vertical feet, before it reaches the buried confluence of the Colorado River (now under man-made Lake Powell), near the foot of Nonnezoshi. Naat-sin'aan, "Head of the Earth," is found here, too, and it takes the form of 10,346-foot Navajo Mountain; this most sacred mountain, together with the surreal world of Monument Valley, forms the southern rim of San Juan Canyon.

Beyond, far to the north of the Navajos' ancestral ground, called Dinetah, is the Dolores River. Like the muddy San Juan, the Dolores is also born in the alpine heights of southwestern Colorado; specifically, the 14,000-foot San Juan Mountains. The Dolores River trundles into the Colorado River 237 river miles above the San Juan confluence near a storied land much like Monument Valley. Here, in a dreamscape of Entrada and Navajo sandstone, 1,500 arches have formed during 100 million years of whistling winds. They bear the appellations of the non-Indians who administer Arches National Park, not the hallowed names of Navajo deities and visions that still adorn the stones and mesas of Monument Valley.

Tumbling into the Colorado River from the west are both the Dirty Devil and Escalante rivers; named by Maj. John Wesley Powell, the Dirty Devil emanates from the 11,000-foot Wasatch Plateau and runs intermittently throughout its tenacious quest to meet the Colorado River at river mile 170. The Escalante River is born of the 10,000-foot Aquarius Plateau and has cut the region of endless canyons that form the great Escalante drainage. Unfortunately, both the Escalante and the Dirty Devil now merge with a dead river that has silenced forever a great canyon once called the Glen, that has since remained the "place no one knew."

Draining the heart of this region is the Colorado River and its principal tributary of the Green River; no other tributary of the Colorado travels as far, or is as famous, as the Green. Fed by glacial runoff of Wyoming's 13,192-foot Wind River Mountains, the Green River travels 730 miles and drops nearly 10,000 vertical feet, before it links up with the Colorado River in the heart of Utah's canyon lands; in so doing, it has cut the chasms of Lodore,

Desolation, Gray, Labyrinth, and Stillwater—all named by Major Powell during his epic 1869 descent—by the time it reaches its confluence with the Colorado River. Together, these two rivers form the heart of Canyonlands National Park before crashing through the huge drops and rumbling through the depths of Cataract Canyon.

Few have ever succeeded at scratching out more than an existence in what remains a haunting region of mystic canyons and naked stone that man could have never imagined if he hadn't first seen it. Only the Anasazi and Fremont peoples are said to have survived, by hunting and gathering, in this rimrock desert of the Great Basin before abandoning it altogether circa A.D. 1300. Some have speculated that drought forced them out; others think the Anasazi and Fremont were being crowded out by hostile tribes—leaving little more behind other than magnificent stone dwellings the white man has encircled with monuments, and an array of figures and messages scrawled in stone the whites are still scratching their heads over.

Only the Utes are known to have existed in the sere, hostile environment that would repulse the hordes of non-Indians who eventually displaced them. Loosely divided into eleven nomadic bands who roamed throughout a 130,000-square-mile ancestral region between the Rocky Mountains and the Oquirrh Mountains on a lifelong quest for food and shelter, the Weeminuche and the Western Ute bands are thought to have inhabited the canyon country and mountain islands of southeastern Utah. Using both poison arrows and deadfalls, the Weeminuche brought down deer, antelope, and elk; they also ate rattlesnakes, chuckawallas, and horned toads; and when things really got lean, they choked down ants with wriggling mouthfuls of crickets and cicadas. Like most other Ute bands, the Weeminuche also harvested piñón nuts in the fall, and when a harvest was particularly abundant, they set up seasonal camps from which they could also hunt deer until the snow flew. Some families were said to plant maize, while most others were known to eat the hearts of mescal plants (as well as paint themselves with the dye-tipped thorns of cactus). Seemingly, the only time food was shunned in this harsh, high desert was during pregnancy; "Certain animals whose 'power' was strong should not be eaten by pregnant women nor hunted by their husbands, for fear the baby might be killed. Mountain lion and the badger, which was thought to be a former Ute shaman, were taboo; also bob-cat and fox" (Callaway, Janetski, and Stewart, 1986, p. 350).

By 1861, the curtain had come down on the Weeminuche band the same way it had on the Taviwach band of the Black Canyon of the Gunnison region; the Weeminuche were forcibly relocated. Those Weeminuche who weren't marched south to the forlorn Ute Mountain Reservation were ordered north to Utah's Uinta Reservation, a land so forbidding that one

OVERLEAF: *Delicate Arch, Arches National Park, Utah.*

of Brigham Young's own surveyors described it as "one vast 'contiguity of waste' and measureably valueless, except for nomadic purposes, hunting grounds for Indians and to hold the world together" (Callaway, Janetski, and Stewart, 1986, p. 356).

It held the same world together that Domínguez and Escalante completely encircled in 1776 during their epic—yet failed—expedition to pioneer a new route from Sante Fe, New Mexico to Monterey, California. As with canyons throughout the West, prospectors, trappers, and surveyors have also probed the region, leaving behind a storied legacy of exploration, science, and greed. The simple petroglyph "1836 D. Julien" remains to this day the most mysterious entry into the history of this vast sweep of labyrinths. Who was he, and how could he etch his stone calling card in Stillwater, Labyrinth, and Cataract canyons thirty-three years before Major Powell made what was believed the first descent of those canyons via the Green and Colorado rivers? One of Powell's own men, Frederick S. Dellenbaugh, thought D. Julien could have been a Spanish padre and even wrote the Vatican, whose assurances laid his theory to rest. Nearly a century later, however, historian Charles Kelly did a little gumshoeing of his own and narrowed the possibilities down to D. Julien being either a French Canadian trapper or Indian trader attached to Antoine Robidoux, who trapped the region during the 1830s and left his own petroglyph etched in stone along the Ute Trail: *"Antoine Robidoux passe ici L. 3/4 E. 13 Novembre, 1837, pour etablire maison traitte a la vert ou Whyte"* ("Antoine Robidoux passed this way on November 13, 1837, to establish a trading house on the green or white river"). Kelly further speculates that D. Julien may have been the first non-Indian to drown in Cataract Canyon, because his last petroglyph was found in the lower end of Cataract Canyon and, because of its position, Kelly believes it may have been made from a boat at high water. No other petroglyphs of Julien's were found below this point; yet all others lead downstream to it.

But whether the mysterious D. Julien was the first to run, or possibly drown in, the big drops of Cataract Canyon will never be known. Consequently, it can probably be safely said that the length of this region wasn't completely traversed by a non-Indian until Major Powell launched his historic descent of the Green and Colorado rivers from Green River, Wyoming, on May 24, 1869. Thus, it falls on the shoulders of Powell, perhaps more than any other early explorer, to describe this canyon region in words that can still be understood today. Of the view from Cataract Canyon, he wrote,

> The landscape everywhere away from the river is of rock, a pavement of rock with cliffs of rock, tables of rock, plateaus of rock, terraces of rock, crags of rock, buttes of rock, ten-thousand strangely carved forms; rocks everywhere, and no vegetation, no soil, no sand. In long gentle curves the river winds about these rocks.

Fremont Cottonwood, Indian Creek, Utah.

Indians, Spaniards, and Anglo pioneers have etched their signs on Newspaper Rock, Indian Creek, Utah.

When speaking of them, [Powell continued] we must not conceive of piles of bowlders or heaps of fragments, but a whole landscape of naked rock with giant forms carved on it, cathedral-shaped buttes towering hundreds or thousands of feet, cliffs that cannot be scaled, and canyon walls that make the river shrink into insignificance, with vast hollow domes and tall pinnacles, and shafts set on the verge overhead, and all the rocks, tinted with buff, gray, red, brown, and chocolate, never lichened, never moss-covered, but bare, and sometimes even polished. Strange, indeed, is *Toom-pin Woo-near Too-weap,* (the Ute words for "Land of Standing Rocks") (Powell, pp. 24–25).

It's strange, indeed, then that except for the Navajo Tribal Park of Monument Valley, the incredible ancestral lands of the Western Ute and Weeminuche are ghostly devoid of both its native peoples and the names they once used to describe the region's most magnificent features. Standing on the rim of Deadhorse Point, watching the cold blue light of dawn wash across their dark canyon home, you yearn to know how the native peoples saw the same river Powell boated, 2,000 feet below. What did they call the hallowed ground that now comprises 527 square miles of Canyonlands National Park? The feeling is like standing on a lonely cliff above Delicate Arch—what did they call this span of stone that frames a rising sun burning over the 12,780-foot La Sal Mountains? What did they call all the other scientifically catalogued arches that now form modern Arches National Park? And what did the names mean? What ceremonies were held beneath them, the same way the Navajo once conducted the Protection Way ceremony beneath Nonnezoshi before it was fenced off and desecrated by the *belogona,* "white man"? We will never know. We never thought to ask before driving them from what must have surely been sacred ground. What is even more disheartening is that there are no native peoples interpreting this wildly imaginative scenery for the hordes of visitors who now hurry through these "parklands" each day.

Fifty miles south, the story is no different. There are no Utes to help decipher the incredible panel of petroglyphs that have been etched on the signpost of "Newspaper Rock" by Indians, Spaniards, and Anglo pioneers who followed the ancient migration route along Indian Creek from the desert reaches of the Colorado River to the forested heights of the 11,360-foot Abajo Mountains for the last 2,000 years. The only native people who still reside at this state park, now located on the fast track to the Needles District of Canyonlands National Park, are the Utes' neighbors from the south—the Navajo. They do so only temporarily, to sell native arts and crafts to hundreds of tourists who come to make home videos of the great rock every week. And while rock art clubs from all over the West are welcome to study the great panel, along with the perplexing array of other "sites" in the area, these native people have been asked to give up their only toehold in this ancestral region. They weren't told why; only that this is their last season to sell handmade arts and crafts to the tourists. It probably has to do with profit. The Grand Canyon's native peoples were ousted from its scenic vistas long ago, undoubtedly because the sales from the South Rim's curio shops were suffering. Perhaps Utah park officials would like to build their own curio shop—and IMAX Theater—here at Newspaper Rock. No other reason can be offered for removing these native people—once again—from their ancestral lands. One elder Navajo woman artist said to me, "How does the government expect our young people to get off welfare—if they take away even this way of living?"

How indeed? Ask that same kind woman what the ancient drawings mean, and she will say, "They are not what the white man thinks they are. The circle is not a wheel; it is something else. But the elders will not even tell the young people, because they have not lived what they have lived; they have not felt what they have felt. So how can the white man pretend to know?"

Non-Indians think largely in terms of science, and when some of the great rock formations are named Navajo, Kayenta, Moenkopi, or the like,

OVERLEAF: *Dawn at Dead Horse Point, Canyonlands National Park, Utah.*

Five young Navajos near Nonnezoshi, *''rainbow rock-span,'' at Navajo Mountain on the Arizona/Utah border.*

little thought is given to what the words actually mean or represent—only that this layer is *X* jillion years old, and that beautiful arches and impenetrable canyons were formed when a mind-numbing array of sterile factors coalesced. No white stopped to think that this land was never ours, or could ever be ours—that it can't possibly be "owned." Once, long ago, a great Blackfoot chief said

> As long as the sun shines and waters flow, this land will be here to give life to men and animals. We cannot sell the lives of men and animals; therefore we cannot sell this land. It was put here for us by the Great Spirit and we cannot sell it because it does not belong to us (McLuhan, 1971, p. 53).

But no one listened.

Yet even when someone now believes they "own" the land, they don't stop to think that the land, the vision of it, and all living things that sustained themselves from it, were once seen through the eyes of native people the old Navajo woman artist called the "Skinwalkers." Instead, interesting pastimes are invented to "experience" the land. We raft the big drops in Cataract Canyon that challenged Powell and possibly took the life of D. Julien; we hike where we're permitted to hike, once the punk-

Mohawk-cropped ranger has sanctimoniously instructed us to watch the official park video before she will issue us a permit to tread "her" ground. We even hang glide and bungee-jump into its mesmerizing depths when all else fails to convey the magnitude of the region. But mostly, we rush about in little vehicles and lumbering road whales, along ribbons of blacktop that have done as much to strip-mine the sanctity of this native ground as the uranium, coal, and water miners have done.

But one only needs to follow the eagle's wing beat south, to Non- nezoshi, to see how native peoples once revered this hallowed ground. It was said that Rainbow Bridge—"rainbow rock-span"—is sacred to the Dinéh, the Navajo people, the same way most others revere their own deities. And before it was made into a National Monument, sacred cere- monials were still held there. In the words of Navajo medicine man Lamar Bedonie,

> Whenever there is a meeting, I talk about this [concern for] Rainbow Bridge. The Navajo people do not like it. Our ceremonies have become difficult. Our prayers and songs are hard to perform. The Nosleep ceremony has become difficult. Now the water [Lake Powell] holds us back. And at that place people are drinking beverages containing liquor. Tin cans and bottles litter the place. We never thought of it in that [profane] way. But it is like that now. When plans were made for [Glen Canyon] dam, we were told it would not be that way (Annerino, 1982, pp. 7–8).

Beyond the world of Nonnezoshi, there exists a whole natural world of sacred mountains, mesas, buttes, rocks, lakes, and streams scattered throughout this awesome tract of redrock monoliths, mountain islands, and earth-ripping canyons that are known only to a handful of medicine men because they are sacred. In her book *Navajo Sacred Places*, Edith L. Watson wrote that these "places [are] where Divinity is closest" and were "before the White succeeded in planting their thorny crops of doubt and disbelief . . . sacred places just as sure as they [the Dinéh] knew the blowing wind and the taste of cold, pure water."

There is something in that—the sanctity of the earth, the natural and mythical life-forms it once sustained. But, thus far, we've missed the point—to see the land for what it once was and might be—if we stopped thinking solely in terms of churning tourodollars out of scenic values, breeding Ticketron reservations out of official park videos, and trading natural vision for satellite-equipped road whales that blot out the eagle's sky trail and the coyote's sandy path. We are the last generation to try bridging the gap between the "Old Ones" and a future that is already upon us; we need native peoples to show the way of understanding the nature of the canyon lands of the High Plateau. The grim blanket of smog that frequently suffocates it, the toxic wastes that now poison it, and the burgeoning crowds that keep loving it the only way they know how are all proof enough that we never saw the land for what it always was. Sacred.

Black Canyon of the Gunnison
Colorado

Here in the mighty gorge of the Black Canyon of the Gunnison
roared a rush of waters maddening in their suggestion of power
and magic. Up and over those mountain walls a parched mesa
flung its story of thirst to the winds that whipped it drier still.

MAE LACY BAGGS
(1918)

The parched winds still whisper to the magic waters, and the canyon still listens, as it has for the last 2 million years. That's how long it has taken the roiling snowmelt of the Gunnison River to carve the somber gorge of the Black Canyon, one foot every 2,000 years, through the ancestral Uncompahgre Highlands first laid down sixty million years ago. Hard to imagine in relation to a single lifetime and a fearful thing, too; man has always feared Colorado's gaping black chasm. It was a place to avoid—maybe look into, but never to enter.

Roaring between the cusp of the Colorado Plateau to the west and the Rocky Mountains to the east, the frigid waters of the Gunnison River were born in the Stone Age glaciers of the Continental Divide, among the 14,000-foot-high peaks of the Sawatch Range and the Elk Mountains— the peaks that form the "Roof of the Rockies." This mountain river and the turgid tributaries of Cochetopa Creek, Lake Creek, and Cebolla Creek, pouring in from the south, and Ohio Creek, Smith's Fork, and the North

15

Fork, emptying in from the north, drain a 4,000-square-mile area and tumble ten thousand vertical feet in a headlong rush to meet the Colorado River 200 miles below.

Once known as the Tomichi to Ute Indians, and the Río de San Xavier to Fray Francisco Atanasio Domínguez and Fray Silvestre Vélez de Escalante, the Gunnison River cut a hard rock canyon that men have held in awe ever since they first walked the earth. Little more than fifty miles long, the Black Canyon drops 2,150 vertical feet from its head near Blue Mesa to its confluence with the North Fork near Austin. That's five times the vertical drop of the Colorado River in the Grand Canyon, with its average 8½-foot drop per mile. The twelve-mile-long, trenchlike gorge forms the heart of the Black Canyon where the Gunnison River cascades nearly a hundred feet down per mile; in the process, it has ground its way like a hydra-headed diamond bit through the Precambrian schists and granites of the Mesa Inclinado. Nearly 3,000 feet deep at Green Mountain, the Black Canyon reaches its most daunting proportions at the Narrows, a 1,725-foot-deep gash in the earth that forms a 44-foot-wide gateway for its raging waters and an 1,150-foot-wide skylight for those peering up from below. Little more than two miles downstream from the Narrows, however, one of the steepest rivers in North America has also created the 2,300-foot-high Painted Wall, a schist-streaked shield of somber gray granite that forms the highest, most imposing wall in Colorado.

Faced with such haunting steepness, and few natural lines of weakness that penetrate its immense precipices, few people were ever willing to tread the fearsome depths of the Black Canyon. That included the Southern Numic branch of Uto-Aztecan-speaking Ute Indians, who first inhabited the region circa A.D. 1000. Eleven Ute bands roamed their 130,000-square-mile ancestral region between the Colorado Rockies and Utah's Oquirrh Mountains; the Taviwach band inhabited the region of the canyon they called Umaweap. They "reportedly felt that anyone going downstream [along the Gunnison River] would never get out alive." They were probably right about that; the only place the Taviwach are known to have crossed the Black Canyon was via a lowly tributary called Red Rock Canyon near the mouth of the gorge. How they lived in Umaweap Cañón, if they traversed its declivitous, boulder-choked, flood-scoured depths, or if they ever used access routes between the North and South rims in the heart of the gorge, cannot be said with any certainty. What ethnographers *can* tell us about their ancient culture was that they were "game rich" compared to the Weeminuche band, which roamed the burned-out-looking mesas and canyon lands of the Colorado River to the west. Elk were hunted in deep snow; deer were run down on horseback; mountain sheep were stalked in the sierra; and both buffalo and antelope were driven over cliffs.

It has taken the Gunnison River two million years to carve the Black Canyon.

If the Black Canyon was ever used as a buffalo jump, it would have been an incredible sight to behold: hundreds of hairy beasts hurtling into the chasm 2,000 vertical feet below. It would have been gruesome, back-breaking work for the Taviwach to haul the heads, hides, and meat out of the deep gorge. They would have had to establish camps near the Tomichi so that the heavy hides could be cured and tanned and long, thin strips of meat could be jerked in the sun. But no archaeological evidence suggests such a scenario.

Knowing the Utes' hunting prowess with large mammals, however, you can't help but wonder, while staring into the Black Canyon, if a small band of hungry Taviwach ever drove a lone bison over the edge and, working together, packed it out to their camps on the 8,000-foot-high rims of Vernal Mesa and Mesa Inclinado. Here they established seasonal camps, among many other locations scattered throughout the high mesa and river valleys on the west slope of the Continental Divide; here they conducted communal rabbit drives, collected nuts and berries, and largely prospered as a culture until the 1860s. Then a black shadow passed over their land, and they were forced to march down a long trail of broken treaties. Ancestral lands, once totaling 56 million acres in Colorado alone, dwindled to 18 million acres in 1868, the first time they were lied to. The second time the Great White Father spoke with a forked tongue, in 1934, they were clinging to less than a million acres. But there was no turning back for these native people; repeatedly displaced from their ancestral lands and forcibly relocated, their culture spiraled toward oblivion. Their natural way of life died, and their dwindling population was eclipsed by the dark shadow they continued to watch pass over them.

Today, sadly, all that remains of the Taviwach band of the Black Canyon is the Ute Indian Museum in nearby Montrose, and it's little more than a forlorn epitaph for a once-great people who were ordered to call the sun-scorched, wind-battered mesa and sagebrush desert of southwesternmost Colorado—in what was once the ancestral territory of the Weeminuche—their home, too.

The Black Canyon was avoided by Domínguez and Escalante as early as 1776, and only studied from the North Rim by the 1873 F. V. Hayden survey. Not until the winter of 1882 did the first non-Indians attempt to traverse the depths of the Black Canyon. Working for the Uncompahgre Extension of the Denver and Rio Grande Railroad, Byron H. Bryant was assigned to survey the Black Canyon from Cimarron to Delta, the Big Deep where no white man had ever been known to tread before. Outfitted for twenty days, the once-eager surveyors spent over two months struggling down the icy, boulder-choked streamed with rods and chains. One surveyor, wrote Richard G. Beidleman (1959, p. 190), was so "good at working on the ice, so fearless that he often had to be restrained from taking chances." As nimble-footed as this expedition proved to be, however, the

The Black Canyon's 2,300-foot-high Painted Wall is a schist-streaked shield of granite.

unforgiving depths of the Black Canyon proved to them that the idea of laying a railroad through it was ludicrous at best. What they did discover from their danger-ridden toils, though, was that perhaps the unchecked waters from the Gunnison River could be diverted over Vernal Mesa for farmers and ranchers in the Uncompahgre Valley. That's about when the idea for the Gunnison Tunnel was hatched. Before that monumental pipedream would become a reality in 1909, however, surveyors would have to complete the work abandoned by Bryant's expedition.

Next in line to match their wits and physical prowess with the Black Canyon was a five-man expedition led by William W. Torrence. Equipped with two canvas-covered oak boats, and "tins of meat, vegetables, and hard tack sufficient for a month's trip," expedition members estimated it would take five days at the outside to traverse the canyon. A month later, the half-starved, bone-weary crew gave up their exploration and abandoned their remaining boat in the Narrows, at the "Falls of Sorrows," before climbing out to the North Rim.

William Torrence later became known as the Father of the Gunnison Tunnel, but the Black Canyon was first traversed after Denver officials sent Geological Survey hydrographer A. Lincoln Fellows a telegram: "Advise me if it is possible to divert Gunnison [River] to Uncompahgre Valley by tunnel under Vernal Mesa?"

Fellows was only too eager to find out, and the first man to throw in with him was none other than William Torrence. Whereas Torrence's expedition was crippled by Sisyphean labors of trying to drag, portage, and carry two heavy wooden boats down the gorge, the idea behind Fellows's 1901 expedition was to swim whatever part of the canyon the members couldn't float on air mattresses. They were in for the adventure of their lives. After blindly swimming the same Falls of Sorrows that had defeated Torrence's 1900 expedition, the haggard pair was forced to replenish their dwindling provisions by cornering a Bighorn sheep and stabbing it to death. Torrence and Fellows not only survived by such tactics, they emerged victorious from the Black Canyon only seven days after they'd first entered it. As Richard G. Beidleman (1959, p. 199) wrote half a century later,

> It is difficult for one, looking down from the rim, to visualize the immensity of the hazards these surveyors encountered. Through this forbidding gorge they had gone, by luck without any mishaps, half-swimming, half-wading, hanging onto their raft, sometimes even lashed to it, pushing and pulling as the occasion demanded, traveling as little as 20 yards in five hours. At night they would seek out a dry ledge above water, sometimes so narrow they had to take turns sleeping on it.

Other parties later attempted to duplicate the traverse: in 1916 legendary Grand Canyon boatmen Emery and Ellsworth Kolb were defeated when their boat was destroyed. In 1934, however, a group of college students, equipped with little more than inner tubes, succeeded. But perhaps no Black Canyon adventurers matched the tenacity and ingenuity of Torrence and Fellows as did a party of climbers who started up the intimidating Dragon Route of the Painted Wall in 1973. Led by Rusty Baillie—renowned for having already climbed the North Face of the Eiger in Switzerland and Norway's ice-plastered, largely overhanging mile-high Troll Wall—the four Arizonans endured nine days of desperate climbing before they stood atop the Painted Wall's 7,922-foot Serpent Point.

Standing on the South Rim at Dragon View, it's difficult to appreciate the achievement of men like Baillie, Dave Lovejoy, Scott Baxter, and Karl Karstrom, who completed such a formidable route—and of Torrence and Fellows, who preceded them through the depths of the great chasm at the foot of the Painted Wall seventy years earlier. But if you stand at that rimbound vista long enough, you will be faced with one of two choices: you will either have to go into the canyon, or you will simply have to turn your back on it. There's no middle ground. In the October 1, 1934, edition of the *Montrose Daily Press,* Rev. Mark Warner wrote, "The Black Canyon of the Gunnison has always held a strange fascination for those who have had the privilege of peering into its awful depths. . . . One will never have seen the Black Canyon in its more majestic and thrilling aspects until he sees it from the bottom" (in Beidleman, 1963, p. 179).

That's even more true today. Although it is now rimmed by a narrow ribbon of blacktop and you can motor up to nine different scenic vistas on the South Rim, you still must descend into the Black Canyon to feel its lasting power. Otherwise, the head-spinning view remains little more than a drive-through postcard; easily seen, readily forgotten once you've driven to the next National Park vista or stepped out of the last IMAX Theater. Yet even today there are no trails that go into the canyon; they're avalanche chutes of loose scree, and the easiest among them plummets an ankle-banging, knee-wrenching, hip-socket-pounding 2,000 vertical feet in a single mile. How the Park Service, on one hand, can lay claim to preserving the natural integrity of lands it stole from the Utes by paving roads and scenic vistas out to the very lip of this great precipice, while refusing to build a safe trail—any trail—into the Black Canyon for less stalwart visitors, is a bureaucratic mystery not easily understood.

If you do venture to the bottom, you will be strolling off a rim flecked by Indian paintbrush and laced with *piñón* and juniper, drop through vertical stands of Douglas-fir, aspen, and spruce, and, by the time you've reached the roaring Gunnison, you'll stand among small clusters of Ponderosa pine and box elder. The waters will run wild at your feet with brown, brook, and rainbow trout; while golden eagles, red-tailed hawks, and white-throated swifts soar high overhead from cliff to cliff. The world will have fallen away, and you will be alone to feel the pulse of the great river that carved the Black Canyon of the Gunnison.

The Canyons of Zion
Utah

We crossed the divide . . . and began our descent into and upon
one of the most scenic portions of America . . . . I do not believe
there is anything on the globe like the canyon of the Rio Virgin.
WESLEY KING
(November 12, 1911)

There isn't. And there isn't anything on earth quite like the 400,000-
square-mile Great Basin sink it's hidden in. Clutched between the Lower
Forty-Eight's two greatest mountain chains, this "sagebrush ocean" laps at
the Sierra Nevada wall running the length of the Pacific Coast to the west
and drifts against the continent-wide Rocky Mountains fencing it off on the
east. Buffered on the north by the tempestuous desert fringe of Oregon,
Idaho, and a corner of Wyoming, the Great Basin is rimmed on the south by
the cliff-warped terraces, mesas, and plateaus that stair-step their way
toward the depths of the Grand Canyon. Here, in the southern reaches of
the Great Basin, where the 6,000-foot-high Colorado Plateau slides off the
rim and into the desert, are found the canyons of Zion.

Corralled on all sides by the colossal land-forms of Hurricane Cliffs,
Paunsaugunt Plateau, and the Vermillion Cliffs, the story of Zion's canyons
is drawn down from the summer monsoons and spring snowmelt that
pours off the 9,000-foot-high Markagunt Plateau. Here, on its forest-
cloaked southern rim, the Río Virgen, "Virgin River," was born.

23

But it all started with a single drop of water; it always has—a tiny dollop hurtling out of lightning-streaked midnight skies; the opalescent tears of melting ice drip, drip, dripping from its crystalline fangs rooted in the bare rimrock. The mission is always the same—search out the sea. Here, in the alpine heights of Zion's 8,500-foot Kolob Plateau, where Douglas-fir and stands of quaking aspen sway in the brisk mountain breeze, minuscule droplets gather, then tumble and swirl to form a million nameless rivulets feeding the torrential creeks and streams that knife their way through its rainbow of stone.

But the Río Virgen did not act alone in its headlong rush to carve its great chasms, en route to joining the Colorado River (now silenced by Lake Mead) 160 miles below. Fed by thundering cataracts of whitewater running wild down its steepest tributaries, the Río Virgen plummets 7,800 vertical feet. Throughout its tumultuous course it gnaws its way through a marbled platform of sedimentary rock laid down before man first roamed the earth. First the ancient flood-born red muds of the Temple Cap formation had to be penetrated; then came the wind-whipped dunes of Navajo sandstone, followed by the dinosaur-tracked red sands of Kayenta, and the sun-baked mortar of the Moenave. It gave way to the petrified wood–pocked Chinle, before rooting through the floodplain detritus of the Moenkopi. But it was in the 2,000-foot-thick Navajo sandstone, a formation that geologists tell us covers 150,000 square miles of the Great Basin, that Zion's most breathtaking canyons were carved long, *long* ago. Some say they were carved 240 million years before local ranchers such as John Winder started wresting a living from the fertile Virgin River valley near the foot of the Moenkopi, when hard-working neighbors paid their respects by saying he was "tough as boiled owl."

But we're jumping ahead of ourselves. Fed by Goose Creek, Kolob Creek, and Deep Creek, the North Fork of the Virgin River drops 76 feet per mile (almost ten times the vertical drop the Colorado River makes through the Grand Canyon), and in so doing has cut the area's most famous canyon. Sixteen miles long, and nearly a half-mile deep, the main stem of the Río Virgen created I-u-goone, or Zion Canyon. The East Fork of the Virgin River, however, drops over a hundred feet per mile as it charges through the 1,700-foot-deep Parunuweap Canyon; that's about how quickly the 2,000-foot-deep Orderville Canyon drops before it trundles into the North Fork a dozen miles above Parunuweap's confluence with it. About midway between Orderville and Parunuweap canyons, however, Pine Creek plummets 300 feet per mile where it plays tag with the Zion–Mt. Carmel Highway. But no other tributary makes the staggering 500-foot-per-mile dive that Taylor Creek does. Pinched off in the northwest corner of Zion National Park, Taylor Creek hunts its way toward Ask Creek, a tributary of the Río Virgen, and, in so doing, has created the enchanting 1,600-foot-deep Finger Canyons of the Kolob.

But the 229-square-mile Zion National Park wasn't officially decreed a park on July 31, 1909, solely for its deep rainbow-hued canyons. The Río Virgen, and the capillary of dramatic tributaries that feed it, also carved the towering stone monuments that thrust themselves heavenward out of the lush valley floor. Named for both Mormon deities and Southern Paiute spirits, the Altar of Sacrifice, the Great White Throne, and the Temple of Sinawava, are just a few of the soaring stone pillars that stand sentinel over anyone who has ever approached Zion Canyon.

One of the first people to gain a toehold in this mysterious world of stone were the Anasazi, or "Old Ones." They wandered the dark recesses of the Kolob Plateau circa A.D. 500 but disappeared about A.D. 1200 without leaving much hard evidence as to why. Was it the Narrows, as one band of Paiute Indians feared? No telling. The Parrusit were one of sixteen bands of Uto-Aztecan-speaking Southern Paiutes, and they followed the Anasazi's abandoned path back toward the terrible canyons not long after the Old Ones vanished. The Parrusit would not tread these forbidden canyons after dark, saying they were inhabited by the spirit beings of the wolf god Shin-na'-wav. The Parrusit also had their own names for the watery tombs in which they feared spirits like Wai-no-pits would seal them. I-u-goone was the name they knew Zion Canyon by, because they viewed the gorge as being "like an arrow quiver." One way in, one way out—along the mysterious defile formed by Mukuntuweap, the name they gave the stream boiling through I-u-goone, because the North Fork of the Virgin River was muk-unt-, a "straight," -u-weap, "canyon stream." Hunters, gatherers, and seasonal agronomists, the Parrusit tilled small plots of corn and squash, gathered pine nuts and *tuna* (prickly pear), and supplemented such staples with meat whenever they were cunning enough to draw down on a deer, or even Desert bighorn sheep that roamed the slickrock before vanishing from the Kolob Plateau during the 1950s. Historian Angus M. Woodbury wrote that the Parrusit lathered the tips of their reed arrows with poison made "by inducing a rattlesnake to bite into a piece of liver, letting it stand a few days and then mixing it with crushed black widow spiders." A potion grim enough to bring any mammal or foe to its knees heaving bile and blood, you'd think. But arrows dripping with snake spit were not enough to stave off the relentless depredations of strangers and slave traders. Rarely numbering more than a thousand strong, the children of the Parrusit band were hunted like dogs by American trappers, Ute and Navajo Indians, and Spaniards during the 1850s. That was about when trader Thomas Farnham reported the children were tracked down and kidnapped during "the spring of the year when young and helpless . . . and when taken are fattened, carried to Santa Fe and sold as slaves. 'A likely girl' in her teens brings oftentimes 60–80 pounds."

Those fortunate enough to elude the vile slavers by slipping into the shadow world of I-u-goone and nearby Parunuweap, "water running into a deep hole," were exterminated with their kin when white settlers and miners traded them disease and famine for their rich ancestral lands. However, one Parrusit man was said to have escaped the apocalypse; by the adopted name of Peter Harrison he lived out the rest of his natural days, until 1945, by running with the neighboring Shivwits band of Paiute who wandered the desolate rimrock of the Western Grand Canyon to the south.

If Native Americans now justifiably lay the blame on Columbus for the coming of the white man, the Southern Paiute would have done well to aim their poison-tipped arrows at Jedediah Strong Smith for causing their population collapse. Smith was believed to be the first non-Indian to actually cross the Parrusit's stony domain when he traversed the Zion region in 1826 and opened the Spanish Trail along the Virgin River between Santa Fe and Los Angeles. That's when the legendary mountain man embarked on an epic journey of exploration that took him from the Great Salt Lake, across the burning wastes of the lizard-eating Mojave, to El Presidio de Nuestra Señora de los Angeles; there Smith and his party of half-starved survivors followed the hard rock spine of the Sierra Nevada north before finally turning back east to cross the Great Salt Desert. Most historians believed Smith followed the Río Virgen across the south-westernmost corner of Utah, but at what point he coursed along the "Adams River" is difficult to discern from his July 12, 1827, letter to Gen. William Clark. Smith's report to the superintendent of Indian Affairs is so vague about this leg of his god-awful route, only one thing remained certain in his mind at the end: "When we arrived at the Salt Lake, we had but one horse & one mule remaining, which were so poor, that they could scarce carry the little camp-equipage which I had along.—the balance of my horses, I was compelled to eat as they gave out."

It wasn't until Maj. John Wesley Powell traversed the region in 1870, en route to pow-wow with the Shivwits Paiute near the Grand Canyon, that the first non-Indians actually explored the depths of Zion's two most famous canyons. Faced with quicksand and deep water, Powell and his men struggled for two days to get down Parunuweap Canyon, and it wasn't until they bottomed out in the Mormon hamlet of Shunesberg, where the East and North forks of the Virgin River form the Río Virgen, that Powell turned his attention toward exploring the narrows of I-u-goone. Traveling like a hunted man, it only took the one-armed explorer a day to "explore this cañón to its head." When he did, Powell wrote, "Everywhere as we went along we found springs bursting out at the foot of the walls." Powell was writing of Zion's "spring line," a natural trestle of hanging gardens, fed by springs and seeps that percolate hundreds of feet through the soft stone. Where this sweet water springs forth, at the contact between the Navajo

Canyoneer John Dell peers over the edge of the Falls, a natural barrier along the North Fork of the Virgin River.

A box elder in Zion Narrows.

and Kayenta formations, lush green bouquets of maidenhair ferns and yellow columbines can be seen bobbing in the misty breeze. Here, at this riparian oasis, Mukuntuweap is laced with fluttering green clusters of box alders, warbling vireos singing from their branches—only one of the 271 species of birds sighted in the region.

Big Springs is also at the heart of Zion's riparian woodland community. Scaly spindace, bluehead suckers, and brown trout swim in the same water that laps at the roots of velvet ash, water birch, and Fremont cottonwoods. Above, the "pygmy forest" of *piñón* and juniper, manzanita, and cliffrose cluster about the rimrock, and craning your neck still higher, the Ponderosa pine belt gives way to a forest of Douglas-fir, white fir, and aspen that grow in Zion's loftiest reaches. What can't be seen from Big Springs any longer are the gray wolf and grizzly bear, hunted to extinction by mountain men and trappers like Jedediah Smith. But if you keep a sharp eye, you may glimpse the wary bighorn—also extirpated from the region before being reintroduced in the Park.

To find the hidden paradise of Big Springs in either Powell's day—or today—the storied, the mythical, the forbidden Narrows had to be traversed, and that was never easy. An undercut, flood-scoured defile, the Narrows was said to be hardly wide enough to ride a good horse through and who really *knew* how high. But word on the dusty streets of Springdale still is that quicksand could swallow you alive. Anytime. And if flash floods struck, God help you! Only the wings of golden eagles, or the spirit beings of the Parrusit, could soar out of its deadly clutches to the forested rimrock high above.

But where the black depths of the Narrows begin, and where they end, is as much a state of mind as a physical realm to penetrate. It goes either way, upstream or down, for those bent on following the same watery canyon path of Major Powell, Jedediah Smith (as some still think), and the Parrusit, who knew to avoid "the canyon after sundown." As most people still seem to do today. Take the dozens of staff-wielding Mormon hikers who can be seen on most summer days following their own path of light from the Mormon strongholds of Ogden and Salt Lake City, down the length of I-u-goone on a ritual, daylong quest through the Narrows to the Valley of Zion. That's where their tenacious, hard-working kin first settled Zion in 1861, years after church leader Brigham Young first heard the call to establish nearby St. George, when it was prophesied that "There will yet be built between those volcanic ridges a city with spires, towers, and steeples; with homes containing many inhabitants."

Converging on the other end of I-u-goone, however, dozens of tour buses, vans, and station wagons can be seen disgorging their own eager pilgrims every summer day, and they too follow their dreams along the same watery path: hundreds of people marching, wading, splashing,

A hiker takes a break during a two-day trek down the length of Zion Narrows.

struggling, swimming, floating, hopping, charging toward Big Springs. Most have poured out of cities and towns across the Southwest, but some have come from as far as New York and China. Most will also use walking sticks to keep from slipping on I-u-goone's moss-covered stones. But, unlike the children of the Mormon angel Moroni who trek down the length of the Narrows between sunup and sunset, most of these good people have come equipped with air mattresses, picnic baskets, inner tubes, boom boxes, styrofoam coolers, and whatever else they can't live without. Just another day at the beach. In Utah.

Year after year, this seasonal migration descends on the forbidden ground of I-u-goone, hundreds of people tromping through the Narrows toward Big Springs every summer day, and now nothing short of a flash flood, or a thundering avalanche of rock that periodically racks this canyon, will stop the relentless tide of human feet surging up the Narrows.

Yet hours later, when everyone emerges from the water-choked black chasm, a soft afternoon light will bathe the Tower of Sinawava. The refracted images of Zion's golden temples will dance in their glassy eyes. But shimmering further beyond, in the distance, the Towers of the Virgin, the Great White Throne—they too will be singing with light. But that is what the people came for. That is what everyone will come for. Light after darkness, in the canyons of Zion.

Canyons of Paria
Utah and Arizona

¡SALISIPUEDES!
"Get out if you can!"
DOMÍNGUEZ AND ESCALANTE
(October 26, 1776)

No one—the native people who tried to survive in it, the strangers who fled or avoided it—described the character of this desolate region and the forces that carved its frightening canyons better than the Spanish padres Fray Francisco Atanasio Domínguez and Fray Silvestre Vélez de Escalante. Enter the region's most spectacular canyon and, yes, you will be awestruck by its sublime beauty, its imposing scale, but one thought will remain with you throughout your mesmerizing journey: "Get out if you can." There are no convenient scenic vistas overlooking the Narrows of Buckskin and Paria canyons. So if you want to see them—the only way you *could* ever see them—you must go into them, on foot. And each time you do, the same tingling fear will haunt you: There's no place to run, and no place to hide if you ever hear the cacophonic clack and thunder of a rampaging flash flood. Overstay your visit, and you *will* hear that deadly music, which has carved these canyons for the last 130 to 200 million years. So you must "Get out *while* you can"—because you are not walking through any kind of canyon you've ever seen or imagined before. You are

33

wading into a deep, dark crack in the earth's crust that at its deepest point is a 3,000-foot drop and at its narrowest is a chest-hugging 2 feet wide. But first you have to find your way into them, subterranean fissures, really, that are penetrated solely by the heavenly shafts of light that illuminate their damp, gloomy depths. And searching these fissures out has never been easy.

Hidden in a forgotten region of towering volcanic peaks, sweeping mesas, and soaring cliffs of raw, blood-red stone, the runoff from the 9,500-foot-high Paunsaugunt Plateau has gouged a sandstone trench so narrow and so deep that on any overcast day you need a flashlight to find the light at the end of its tunnel. That's because it's lost in the land of the giant, the 100,000-square-mile Colorado Plateau, a 6,000-foot-high area that covers a four-corner region of Utah, Colorado, New Mexico, and Arizona. It is a land, wrote geographer Herbert E. Gregory (Gregory and Moore, 1931, p. 12), that is sculpted "on so enormous a scale that features . . . unnoticed here would be prominent and picturesque landmarks in other surroundings."

Gregory was right about that. And there's no better example of his observation than the Kaiparowits Region, a hardscrabble 5,400-square-mile corner of the Colorado Plateau that is covered with its own immense subplateaus that were named for native people such as the Ute and Southern Paiute. Take Paunsaugunt Plateau, Paria's headwaters; its name evolved from Paunsauganti, because it was said to look like the beaver that once swam its watery courses. Or try the 7,000-foot-high Kaiparowits Plateau; its name was taken from Kaivavič; like the 9,000-foot Kaibab Plateau—boiled down from kaiva-viči-ci-nwɨ—because *both* adjoining plateaus were said to resemble "a mountain lying down." These plateaus, drained by rivers and streams named by the Spaniards who fled them, gave way to the world's most famous canyon, the Grand Canyon of the Colorado. But it was a lowly tributary of the Río Colorado, "red river," what the ancient Southern Paiute have always called the Paria-pa, "elk water," that cut what is probably the narrowest and deepest canyon on earth. Then, perhaps that was because Paria-pa takes its own name from a mythical elk seen leaping from a high cliff into its roiling waters.

Trickling off the southern ramparts of Paunsaugunt Plateau, the Paria-pa drains a 935-square-mile tract of the Kaiparowits Region before spewing into the Colorado River 90 miles below. Where the Paria-pa honed its way through the seofan-tiered layer of mud, silt, and sandstones, and created the Narrows of the 40-mile-long Buckskin–Paria canyon system, geologists call them "inclosed meanders"; its erratic course bears a striking resemblance to the loopy S-shaped tracks left by the Western rattlesnake that can still be seen slithering along the sandy rims. This meandering is no more evident than in Paria-pa's search for the Colorado River 6,500 vertical

Canyoneer Richard Nebeker wades through the "Cesspool" during a 40-mile trek down the length of Buckskin Gulch and Paria Canyon on the Arizona/Utah border.

feet below its headwaters in the strange hoodoo world of Bryce Canyon. Instead of following the natural depression of House Rock Valley to reach the Colorado River, though, the Paria-pa did the impossible—it gnawed its way through monolithic buttresses, cockscombs, and walls of stone that might have otherwise prevented it from creating a dark, nearly lifeless fissure in the earth that most men have always tried to avoid.

Like the canyons I-u-goone and Mukuntuweap 70-odd miles west, the Anasazi (the "Old Ones") were also said to inhabit the Buckskin–Paria domain. But except for a few scattered ruins poised near the mouth of the Paria-pa, and the Split Twig figurines now thought to be their effigies for a successful hunt, archaeologists can tell us little else about this lost civilization. Current theory suggests the Old Ones were forced to abandon the region because they either faced a biblical-style drought, or because the Ute and Southern Paiute, migrating south in search of new lands, drove them off circa A.D. 1200. But no one knows for sure.

Nor can ethnographers tell us much more about the vanished bands of Southern Paiute who scratched out a living in the high, barren deserts and deep canyons of the Paria-pa. In fact, the Paria-pa was the natural dividing line between the Kaibab and Kaiparowits bands of Southern Paiute. The Kavavič-ŋɨwɨ, "mountain lying down people," roamed the forested Kaibab Plateau region that comprises most of the North Rim of the Grand Canyon. But heavy winter snows prevented the Kavavič-ŋɨwɨ from living in permanent settlements on the North Rim and, like the Kaiparowits, they moved to lower elevations during winter months to hunt and gather. The Kaiparowits, on the other hand, were known as the Kwaguiuavi-nɨwɨŋ, "seed valley people," and they inhabited the harsher, more desolate canyon and desert region of the Paria River and Kaiparowits Plateau; they called this land asikaivɨ, "all rocks, no trees," perhaps because it accurately described the daily life of anyone trying to eke a living from its stony desert mesas and canyons.

One animal that thrived in the Kaiparowits Region, though, was the Desert bighorn sheep, and the Kwaguiuavi-nɨwɨŋ hunted them in Paria Canyon, either by driving them over cliffs or corralling them on narrow ledges before stoning them to death. Meat from Desert bighorn, antelope, and deer, as well as small game, was cooked on *iyɨmpɨ*, juniper, which also provided warmth for winter camps; while pine nuts, berries, cactus, and seeds proved to be dependable staples when hunting was lean.

However, nothing was as scarce in the parched desert plateau region of Paga, "big water" (the Colorado River), as water itself. And waterholes determined where these nomadic hunters and gatherers lived, as much as the availability of food. Waterholes were not owned, but the names of those families who dwelled in brush huts nearby came to be associated with each waterhole, for example Muavigaipɨ, "Mosquito Man," Nakavaiptinkaipi, "No Ears," and Tumunsokont, "Black Moustache." These were some of the names whispered by the parched lips of roving band members, who sometimes stopped to quench their thirst at these communal waterholes. The Kaiparowits and the Kaibab bands also had names for every physical feature of the desolate region they inhabited before they, too, marched into ecological extinction.

Leading the charge were Domínguez and Escalante. Only the Spanish explorers were so turned around during their epic five-month-long quest to pioneer a route from Santa Fe, New Mexico, to Monterey in Alta California, "Upper California," that by the time they doubled back for Santa Fe, they were nearly defeated by the imposing wall of Echo Cliffs near the confluence of the Paria and Colorado rivers. *"Salisipuedes"* was etched into the face of the region shortly before their Indian guide led them

Buckskin Gulch at one of the few wide spots in this deep, narrow canyon.

through the broken cliffs, and Domínguez and Escalante went on to pioneer the "Crossing of the Fathers" in Glen Canyon.

Mountain men and trappers fared little better during the 1820s. They tended to avoid the blistering deserts and impenetrable canyons, because, as Herbert Gregory wrote, "It was not beaver country nor buffalo range. Little food was to be obtained, and few Indians were there to be robbed. The fur hunters were interested in beavers, not scenery" (Gregory and Moore, 1931, p. 6).

Not until polygamist John D. Lee was ordered by the Mormon church to establish a reliable ford across the Colorado River was Paria Canyon first descended by a non-Indian. Lee did what no man had ever done before, or would ever do again; he drove sixty head of cattle down the length of Paria Canyon in order to establish Lee's Ferry near the mouth of the canyon with his seventeenth wife, Emma. In *A Mormon Chronicle: The Diaries of John D. Lee: 1848–1876,* edited by Juanita Brookes and Robert Cleland (1955, p. 178), Lee described what probably ranks as one of the most incredible cattle drives of all times:

> We concluded to drive down the creek [Paria], which took us Some 8 days of toil, fatuige, & labour, through brush, water, ice, & quicksand & some time passing through narrow chasms with perpendicular Bluffs on both sides, some 3,000 feet high, & without seeing the sun for 48 hours, & every day Some of our animals Mired down & had to shoot one cow & leve her there, that we could not get out, & I My Self was under water. Mud & Ice every day. We finally reached within 3 ms. of the mouth with 12 head, leveing the remander to feed some 10 ms. above. 4 days had Elapsed since our provisions had exasted, Save Some beef that we cut off the cow that was mired.

Just six years after this remarkable cattle drive, a firing squad executed Lee for his part in the Mountain Meadows Massacre. Ironically, Lee's own gun had misfired, but he was still pegged as the sole scapegoat for the heinous slaying of 122 wagon-bound emigrants murdered by Mormons and their Paiute allies in 1857.

But if driving a herd of cattle down Paria was strange, making a military reconnaissance of it a year later was probably even stranger. What strategic importance the isolated canyon held for the U.S. military was never spelled out (even the author of the report remains mysterious—the report is only signed T. V. B.). An eleven-man expedition, led by "Lt. W. L. B." of the Army Corps of Engineers, was assigned to explore the "rim of the Great Basin of Southern Utah, thence go to the Colorado River, ascend and explore the cañón of Paria, and return." However, the expedition met with disaster not long after their "Pah-Ute" guide abandoned them, calling them "whitey man big fool." This remark had to do with the fact that expedition members stoked their nightly campfires with a "half a dozen dead pine trees . . . that [lit] up the pine woods for miles."

Unlike most National Parks in the Southwest, there are no convenient scenic vistas overlooking Buckskin Gulch and Paria Canyon.

Nevertheless, they pushed on without their Paiute guide and after three days' struggling through icy waters up the length of Paria, the expedition cook drowned. While leading the expedition through a deep, icy pool, Kittleman fell off his horse as it kicked and shuddered in the frigid water; he disappeared. Thanks to the heroic efforts of "Lt. W. L. B.," Kittleman was dragged out, blue with cold. All attempts to revive him failed, though, and defeated expedition members stuffed his body in a crack, before fleeing Paria Canyon for good.

Emerging from the depths of Paria Canyon today, you enter a world both empty and sublime. Sublime, because it's wonderfully alive with the *piñón*-juniper and Great Basin desert scrub communities that shroud its deepest reaches. Petroglyphs still mark the declivitous routes used by the Kwaguiuavi-nɨwɨŋ, or perhaps even the Anasazi, to get into and out of the canyon from the 6,000-foot Paria Plateau; while rusted machines nearby at Judd Hollow remain monuments to the futile efforts of such

At the bottom of Buckskin Gulch.

pioneer stockmen as John Adams, who dreamed of pumping water out of the Paria to his dying cattle on the sand-warped rim 2,000 feet above. Yet the heart of the Paria-pa is a strangely vacant world, because gone are the native Kavavič-ŋɨwɨ and Kwaguiuavi-nɨwɨŋ who first fed the pulse of the Paria-pa, and gone are the antelope and Desert bighorn they subsisted on.

With the disappearance of these native people, and an ever-growing number of life-forms they subsisted on—here and elsewhere—you can't help but wonder about our very concept of wilderness today: Are these great canyon wildernesses really "natural," or are they still "wild"? In *Touch the Earth* (1971, p. 45), Chief Luther Standing Bear, of the Oglala band of Sioux, shared some insightful thoughts:

> We did not think of the great open plains, the beautiful rolling hills, and winding streams with tangled growth as "wild." Only to the white man was nature a "wilderness" and only to him was the land "infested" with "wild" animals and "savage" people. To us it was tame. Earth was bountiful and we were surrounded with the blessing of the Great Mystery. Not until the hairy

man from the east came and with brutal frenzy heaped injustices upon us and the families we loved was it "wild" for us. When the very animals of the forest began fleeing from his approach, then it was that for us the "Wild West" began.

Imagine, then what it would have been like to journey through the shadows of the Paria-pa with a small band of hunters on a ritual quest 50 miles north to the Table Cliffs. Here in the forested realm of what they knew as Kaiparawɨcɨ, "Mountain Boy Named by Coyote," the desert- and canyon-dwelling Kwaguiuavi-nɨwɨŋ still had names for four different kinds of bears they revered and hunted! Tocakwiacɨ was the white bear, Gasikwiacɨ was the gray and black bear, possibly grizzly, Ontonkwiacɨ was the brown bear, and Ankakwiacɨ was the rare red bear. Envision these stalwart people, armed with little more than rude knives, stone-tipped arrows, and flimsy bows, speaking in reverence to their great bear spirit, Kagun, "Maternal Grandmother," before they let their arrows sing, or were brutally mauled: "Old woman, I am going to kill you. I want to eat your meat. Do not be angry; do not kill me."

I still wonder what that might have been like, on the dark stony path to the Paria-pa.

Canyons of
Redrock–Sacred Mountain
Arizona

Canyons . . . seam and slash the sides of mountains . . . into
the depths of the wildest, tumbledest, most upheaved rock
country the eyes of man ever gazed upon.
GEORGE WHARTON JAMES
Arizona the Wonderland (1917)

And nowhere is that scene more vivid than from this wind-swept perch
high atop the Black Hills of central Arizona. Anvils of towering cumulo-
nimbus can be seen erupting from the summit pyramid of a dead volcano
fifty miles north, while curtains of torrential rain and a holocaust of white
lightning hammer the tangle of canyons fanning out below. This storm-
battered mountain is Huchassahpatch, "Big Rock Mountain," sacred
mountain of the Havasupai. It is also Nuva-teekia-ovi, "the Place of Snow
on the Very Top," sacred mountain of the Hopi. And it is Dok'o'sliid, the
Navajo's "Sacred Mountain of the West." And sometimes it is even San
Francisco Mountains, the highest range in Arizona. By any name, its
12,633-foot-high crest rides defiantly above the 6,000-foot-high Colorado
Plateau. Summer monsoon unleash their subtropical fury against its soaring ramparts and the serpentine canyons that cleave its base. Formed by the
100,000-square-mile Colorado Plateau, the southern rim of this gargan-
tuan mesa slides out from beneath the Plateau province and avalanches
into the Basin and Range province 4,000 feet below. The result is a 200-

43

Children frolic in Oak Creek Canyon's Slide Rock.

mile-long brink called the Mogollon Rim. Here, on the western end of what some have called the Intermontane province, are found the Canyons of the Redrock–Sacred Mountain—carved by the same relentless monsoon rains, and deep snowpack, that have trundled off the subalpine heights of that great sacred mountain in search of the desert and the sea, for the last 400,000 to 1.8 million years.

Among the maze of canyons that have incised Arizona's statewide precipice (named for the governor of New Mexico in 1712, Don Juan

Ignacio de Mogollon [*mogoy-yown*]), none is more rugged than the dry, boulder-choked depths of Sycamore Canyon. Twenty-five miles long, and over 1,500 feet deep, the supernal waters of the San Francisco Mountains plummet another 3,000 feet from the head of Sycamore Canyon into the desert river formed by the Río de Los Reyes, "River of Kings," or Verde River. In so doing, they have nurtured life throughout an alluring mix of biological communities that range from the Canadian forest of Douglas-fir, white fir, and Ponderosa pine near its 6,600-foot upper end, through the deciduous riparian community of ash, box elder, walnut, sycamore, and cottonwood along its gnarly stream bed, to the cactus-studded grasslands of the Upper Sonoran desert near the 3,600-foot level at its Verde River confluence. This "miniature Grand Canyon," however, was created where that same pummeling runoff eroded its way through a sedimentary layer of rock formations that many people continue to associate with the great chasm 65 miles north. Most recognizable among them is the white band of Kaibab limestone, the interbedding of Toroweap, the massive buff-colored walls of Coconino sandstone, the cross- and interbedded red walls of Supai, and the truss of pale gray limestone called the Redwall.

Across the 7,000-foot-high Secret Mountain divide, however, is the West Fork of Oak Creek, and in a word it is simply the region's most beautiful canyon. Little more than twelve miles long and 1,500 feet deep, the lush West Fork of Oak Creek is cloaked with pristine stands of White fir, Douglas-fir, and Ponderosa that poke at the heavens from soaring white aprons of Coconino sandstone that march through it. Near its confluence with the main stem of Oak Creek Canyon, the West Fork is at its most magnificent, because here it joins its mother canyon, which by itself has entrenched a 17-mile-long, 2,500-foot-deep canyon that gives way to the scenic heart of Sedona's Redrock Country: a fairytale mural of rainbow-hued walls, buttresses, mesas, and pinnacles that are more easily imagined in a dream.

The stuff of legend and fantasy, Oak Creek Canyon is also the whimsical dividing line between the West Fork and Sycamore Canyon to the west and Wet Beaver Creek and West Clear Creek to the east. Of these five major canyons that drain into the upper Verde River, Wet Beaver Creek has always been the canyon swimmer's dream—or nightmare. West Clear Creek is nearly twice as deep, with walls looming some 2,000 feet above a cold, clear stream in which rainbow trout swim, versus the 1,000-foot-deep slash that created Wet Beaver Creek and the jungle of poison ivy that follows it. But hunting this canyon from either upstream or down, the native *ʔwipukpáya* and *ndee-nnee* either faced swimming two dozen, chest-heaving plunge pools within its abrupt 14 miles or stretched out their

OVERLEAF: *Sedona's redrock country.*

struggle by swimming the twenty-eight plunge pools that permanently flood West Clear Creek's 30-mile course. Swimming and carrying a deer carcass through either canyon had to have been a nightmare—or a rite of passage.

Among the first to explore the region's canyons on hunting, gathering, and fishing forays were the *sinagua*. Archaeologists tell us this ancient culture developed in the vicinity of the San Francisco Mountains circa A.D. 600. But evidence of their primeval wanderings through the canyons of Redrock–Sacred Mountain wasn't described for the layperson until George Wharton James penned *Arizona the Wonderland* in 1917:

> In the walls of this canyon [Sycamore], hundreds of cliff dwellings may be seen and on some of the salient points are buildings that appear to be fortresses. . . . Some of the dwellings have been excavated and arrow points, spearheads, stone axes, ropes made of fiber of beargrass, or amole, sandals, ears of corn . . . have been collected. (p. 389)

What happened to the *sinagua* after the lowly volcano of Sunset Crater blew out circa A.D. 1066, and choked the region with volcanic ash, nobody can say for sure. But two distinct tribes occupied the forest-shrouded canyons abandoned by the *sinagua* not long after. Of them, the Upper Yuman-speaking Yavapai are easiest to link to these canyons with certainty. Occupying an immense ancestral land totaling some 20,000 square miles, the Yavapai "were not confined to a single ecological area," as ethnographer E. W. Gifford wrote, "but ranged over a wide variety of territory . . . from blistering desert to shady mountain streams, from lower Austral life zone to Canadian life zone" (Annerino, 1991, p. 183). Divided into four principal bands, the *tòlkpáya* roamed the sere, unforgiving western deserts the Spanish allegedly called La Arido Zona, or Arizona; while the kwèkpáya inhabited the central Arizona mountains surrounding the Salt and lower Verde rivers. But it was the northeastern Yavapai who thrived in the canyon country on the western end of the Mogollon Rim. Among them, the *yavpé* inhabited Sycamore Canyon region and the isolated mountains to the south; but it was the *ʔwipukpáya*, "people at the foot of the rocks," who lived in the heart of the area's canyon country, including West Clear Creek, Wet Beaver, Oak Creek, and the West Fork of Oak Creek.

Apparently, it was a region these hunters and gatherers willingly shared with the Northern Tonto, one of five bands of Athabaskan-speaking Western Apache, or *ndee-nnee*, "man, person," or Apache. However, some ethnographers have argued that the Northern Tonto—who the Spaniards called *tonto*, "crazy people" or "without minds," because the Spaniards were ignorant of their language and customs—were actually Yavapai who intermarried with other bands of Western Apache. Whether they were or not, they had the wrong name. The mere word "Apache" struck fear into the hearts and minds of white settlers throughout the territory during the

Fall colors in the depths of the West Fork of Oak Creek.

1800s, as well as the legions of hacks who eagerly embellished—and made up—tales of treachery and death they ascribed to Apaches in order to peddle their dime novels. Unfortunately, the image of "savage Apaches" stuck, and it eventually brought about the unconscionably brutal subjugation of all Arizona's native peoples, including the canyon-dwelling *n*d*ee-n*n*ee, yavpé,* and *ʔwipukpáya.*

Before their cultural demise, however, these native canyoneers lived in a biologically diverse region that offered them meat from Desert bighorn, deer, and antelope, as well as staples such as cactus fruit, walnuts, acorns, and *piñón* nuts. Like many hunting and gathering peoples, the *n*d*ee-n*n*ee, yavpé,* and *ʔwipukpáya* moved seasonally as environment and needs dictated. When they did, their elders counseled them,

> Do not be lazy. Even if there is a deep canyon or a steep place to climb, you must go up it. Thus, it will be easy for you to get deer. . . . When you trail deer you may step on a rock. If the rock falls from under you, you may fall and get hurt. If there is a thick growth of trees ahead of you, don't go in it. There might be a mountain lion in a tree ready to attack you. (Basso, 1983, p. 472)

Canyoneers in Sycamore Canyon, one of Arizona's most rugged canyons.

If mountain lions weren't ready to attack the *yavpé, ʔwipukpáya,* and *ndee-nnee* in their canyon environs, U.S. troops were all too willing. But three centuries before these native people first suffered deprivations at the hands of the white man, they gave "metals in addition to food as a sign of friendship" to the region's first foreign visitors. Lured by the promise of silver in the mountains south of Sycamore Canyon, Capt. Antonio de Espejo led a nine-man expedition out of Zuni, New Mexico, in May 1582 and either descended Sycamore Canyon or Wet Beaver Creek in order to reach the Black Hills. But nobody knows for sure: not Espejo's diarist Diego Pérez de Luxán; not the historians who later linked Espejo to both canyons four centuries afterward. Of their epic quest for precious metal, however, Luxán was certain about one thing when the expedition finally emerged from their tortuous canyon route:

> We descended a slope so steep and dangerous that a mule belonging to Captain Antonio de Espejo fell down and was dashed to pieces. We went down by a ravine so bad and craggy that we descended with difficulty to a fine large river which runs from northwest to southeast. At this place this river is surrounded by an abundance of grapevines, many walnut and other trees. It is a warm land in which there are parrots. (Bartlett, 1942, p. 29)

It was another three hundred years before the next group of non-Indians actually descended one of the Redrock–Secret Mountain canyons. But Gen. W. J. Palmer had a better idea about which canyon they were in. Assigned by the Kansas City Railroad Company in 1867 to survey the most practical route to the Pacific, General Palmer and his men were en route back to the Little Colorado River from Chino Valley when they took a mean shortcut through the rugged depths of Sycamore Canyon—only they were ambushed by "Apache." Wrote Palmer,

> We suddenly heard a shot from the brink of the canyon at our rear, and the dreaded war-whoop burst upon us. Then we looked up to the right and left, ahead and to the rear; but the walls seemed everywhere as tall as a church steeple, with scarcely a foothold from top to base. They had looked high before, and the chasm narrow, but now it seemed as though we were looking up from the bottom of a tin mine, and no buckets to draw us up. Soon the shots were repeated, and the yells were followed by showers of arrows. (Farish, 1918, p. 112)

Incredibly, General Palmer engaged the Indians from his vulnerable position on the floor of Sycamore, and even more incredibly, they escaped with their lives.

The region's native people were not so lucky when Gen. George Crook arrived on the scene and "ordered that all 'roving Apache' were to be on this reservation [Río Verde] by February 15, 1872, or be treated as hostile." Some fled; others fought back; many were slaughtered. In the end, the Yavapai-Apache were offered a paltry 2,000 acres on two reservations they were ordered to call home, and the canyons of the Redrock–Sacred Mountains lost the native people who held them most sacrosanct.

Today, unfortunately, you will not find a tribe of modern *sinagua, yavpé, ʔwipukpáya,* or *ndee-nnee* living peacefully near the mouth of these canyons, as the last band of *ʔwipukpáya* did near Indian Gardens four miles north of Sedona in Oak Creek Canyon. Nor will you see them tending hand-tilled plots of corn, small orchards, and cattle—as the small tribe of Havasupai still do in the western Grand Canyon, and as the vast nation of

Built in 1931 as a line camp for cowboys, Taylor Cabin is a popular destination for canyoneers exploring Sycamore Canyon today.

Tarahumara still does in Las Barrancas del Cobre region of Chihuahua. Instead, you will find the chic, bustling, five-star resort community of Sedona built at the mouth of these great canyons, which is now home to 10,000 dudes, cowboys, ranchers, artists, retirees, movie stars, realtors, and developers. In addition to these proud locals who now hold "their" land sacred, some six million people visit the Redrock Country each year. Of them, perhaps none are more bizarre than the tribe of New Agers who've desecrated sacred Yavapai-Apache ancestral grounds by building replicas of Plains Indian-style "medicine wheels" at the mouths of the area's most spectacular canyons. Long associated with Northern Plains tribes such as the Blackfoot, Crow, Cheyenne, and Sioux, who built stone medicine wheels up to 80 feet in diameter for sacred dances and astronomical observations, there is no cultural, historical, or archaeological evidence that any of Arizona's native peoples ever built or used them. Yet like the Plains people they try to emulate, these born-again "natives" also ride fast ponies—only they call them Black Saab and White Volvo—and they hitch them within easy walking distance to their stone alignments so they can pay homage to the Great Spirit (whose name, in their eyes, has shape-shifted to "Harmonic Convergence"). Only in Arizona have the strangest elements converged to rebuild the London Bridge, the highest water fountain in the world—in one of the driest deserts, a human biosphere that doesn't include one Native American, and now the Harmonic Convergence. Yet not even the local Yavapai can explain the strange ways of these people. As one elderly woman told me when asked about the "medicine wheels" in nearby Sedona, "I don't know *what* those white people are doing over there."

Perhaps the best way to flee the mumbo-jumbo, the smog, the glitter, the din of what National Park officials had envisioned as a National Park gateway before being waved off by irate "locals," is to go into the heart of any of these canyons. Because it's there, you can still find the pulse of its alluring natural world. In *Call of the Canyon,* author Zane Grey wrote, "This canyon of gold-banded black-fringed ramparts, and red-walled mountains, which sloped down to be lost in the purple depths. That was the final proof of the strength of nature to soothe, to clarify, to stabilize the tired and weary and upward gazing soul."

In the canyons of Redrock–Sacred Mountain, that's all man has ever needed.

Grand Canyon of the Colorado River
Arizona

Alone with my campfire, I gaze on the . . . countless campfires
around which are gathered the people of a dying race.
I feel that the life of these children of nature is like the dying day
drawing to its end; only off in the west is the glorious light of
the setting sun, telling us, perhaps, of light after darkness.

EDWARD CURTIS
The North American Indian (1905)

There is light after darkness, and for the native peoples who lived, thrived, and died within its breach, it was the Grand Canyon. For the strangers who marched through their ancestral lands and peered into its unfathomable depths, it was the Grand Canyon. And for those still fleeing the iron grip of modern civilization, it will always be the Grand Canyon. Light after darkness, in a world torn by disarray.

From the cold, tundra-covered summit of the San Francisco Mountains, a great gash in the earth can be seen 70 miles north, and it glows gold, pink, then gray with the setting sun. This is the Grand Canyon of the Colorado River, and nothing else on earth compares to it. It is a natural wonder. It is a sense of place. It is a state of mind. Then, no other canyon on earth was created with the unrelenting force of a river like the Río Colorado, the "Red River." Emanating from the west slope of the Continental Divide in the Colorado Rockies, the river once called the Grand drains

Maidenhair ferns at Stone Creek Falls below the North Rim of the Grand Canyon.

244,000 square miles and drops 10,000 vertical feet by the time it finally trickles through the vast delta where the Grand Canyon empties into the Sea of Cortez. The most famous stretch of this 1,400-mile-long river has always been the 7,000-foot-deep, 277-mile-long chasm gouged through the tier-stepped Colorado Plateau.

Officially consuming 2,000 square miles, the Grand Canyon is far more vast than its tidy bureaucratic borders. And viewing that great chasm from the summit of the San Francisco Mountains, and its 12,633-foot sacred peak, it's easy to imagine how the river first channeled its way through the Kaibab Upwarp of the Colorado Plateau 25 to 65 million years ago. Harder to see in the glimmering twilight are the seventy-odd tributary canyons that drain its North and South rims and comprise the labyrinthine maze of canyons that make the Grand Canyon what it really is. Like the raging mother river these hanging flood-swept canyons have always sought out, their waters have also ripped their way through a kaleidoscopic layer of rock formations that date back from the 250-million-year-old Kaibab limestone that roofs the canyon to the 1.7-*billion*-year-old Vishnu schist found at its floor.

Among the Grand Canyon's largest tributaries is the Little Colorado River Gorge. In the luminescent haze of twilight, it can be seen slithering across the dark plateau like a black serpent hunting its way toward the depths of its canyon home. Fifty-seven miles long and 3,424 feet deep, the Little Colorado River was born of sacred waters that melted from the snowy heights of the *dzil ligai*, "white mountain" over two hundred miles upstream. Sixty-one river miles north of the confluence of the Little Colorado River and its mother river lies Paria Canyon. Like the Little Colorado River Gorge, Paria's own forty-mile-long, 3,000-foot-deep canyon was also trenched by waters born from the snowy highlands of native peoples. The story is the same for Kanab Creek 82 miles downstream from the mouth of Little Colorado River Gorge. Sixty miles long and 3,892 feet deep, Kanab Creek is an immense chasm dwarfed only by the canyon it feeds. So is the 50-mile-long, 3,644-foot-deep drainage of Cataract and Havasu canyons; they spew their turquoise waters into the Colorado River thirteen miles below the mouth of Kanab Creek. This is the Grand Canyon—not one huge gorge, but a Himalaya-sized region of "inverted mountains" that form the image of a single canyon that has nearly always been called *grand*.

The first people to inhabit this sublime domain were said to be the Anasazi, "Old Ones," and the *cohonina*, Hopi for "Havasupai." Hunters, gatherers, and masters of a bioregion that extended from the Canadian forests of the North Rim to the Mojave desert wastes of the inner canyon, both ancient cultures thrived side by side until they disappeared from the

OVERLEAF: *Summer visitors enjoy cocktail hour on the North Rim of the Grand Canyon and superb views of (left to right) Deva, Brahma, and Zoroaster temples.*

A backcountry ranger coordinates a search for a lost hiker.

region circa A.D. 1200. The Anasazi, who roamed the desolate, hard-rock sweep of the Colorado Plateau beyond the Grand Canyon, also dwelled throughout the inner canyon summer and winter. Deer were hunted, *piñon* nuts were gathered, and mescal stalks were roasted in *yanta* ovens in the cool heights of the canyon's rims; while corn, beans, and squash were tilled in its balmy depths near river's edge. Then, *p-ffft!*—they vanished, leaving few clues as to why.

The native peoples who sought out these environs shortly after the Anasazi abandoned the 2,500 stone dwellings still found throughout the Grand Canyon were as diverse a group of tribes as the region has ever supported. Believed to be direct descendants of the Anasazi, the Hopi (*hópi,* "wise, knowing," or *hópʰi,* "good, peaceable") have lived on the dry, wind-swept, sand-dusted mesas sixty miles east of the Grand Canyon at least as far back as A.D. 1150. And to this day, some ten thousand Hopi still dwell on 2.5 million acres of traditional lands encircling Oraibi, the oldest continuously inhabited village in the contiguous United States. Prolific runners who ran for travel, ceremony, and sport, the Hopi hunted deer and rabbits but sustained themselves largely through their masterful efforts at dry and floodwater farming, producing bumper yields of corn, beans, squash, and even orchards on ground that the white man would later disdain. They held a sacred reverence for the Grand Canyon; and it was even more commonly known that the Hopi made ritual treks for salt across the sere reach of the Painted Desert, and thence down the Little Colorado River Gorge to cliffside deposits hidden in the bawling depths of the Grand Canyon.

Two hundred hard miles by foot west of Oraibi, along the ancient trade route that once linked both tribes, live the Havasupai, "People of the blue-green water." Next to the Tarahumara, they are one of the only native peoples still known to dwell within their ancestral canyon environment. Descendants of the ancient Cerbat people, the Yuman-speaking Havasupai have lived in the paradisal narrows of Cataract and Havasu canyons since time immemorial. Some say since A.D. 1200. Like the Hopi, the Havasupai were farmers, but they were also skilled hunters, who tracked down Desert bighorn sheep, which still roam the red rimrock encircling the village of 500 Havasupai.

Still further west along the South Rim of the Grand Canyon, near Diamond Creek, is a neighboring tribe of Yuman-speaking people called the Walapai, "People of the pines." Blood kin to both the Havasupai and the Yavapai, the Walapai once inhabited a chartless realm that extended across the South Rim from the Little Colorado River Gorge all the way to the Lower Colorado River near the western terminus of the Grand Canyon. They too stalked Desert bighorn, deer, and antelope, as well as harvested cactus fruit, mescal, and *piñon* nuts. Like native peoples almost everywhere, the Walapai's traditional culture was also crushed—but first they sought to fight back by adopting the shamanic Ghost Dance. Ethnographers tell us the Walapai had two good reasons for stomping out its haunting beat: to remove the "Anglos from traditional Walapai Territory . . . and [to resurrect] . . . dead ancestors." The Ghost Dance was introduced to them by the Southern Paiute with whom they traded via intercanyon routes that linked the south side of the Colorado River with the north. The Walapai danced their last mournful steps in 1905, when the dance was outlawed, and today live on 1 million acres on the South Rim of the Western Grand Canyon in what, at that time, was officially decreed "the most valueless land on earth for agricultural reasons."

The Southern Paiute did not fare much better. Once freely roaming 11,000 square miles of the Arizona Strip, Paiute bands such as the Shivwits, Uinkaret, Kanab, and Kaibab hunted and gathered on the North Rim's great plateaus by those names since A.D. 1300. But their Ghost Dance died about the same time the Walapai sun set, and today some 200 Kaibab Paiute live on 120,000 bleak acres "behind the Grand Canyon."

Backcountry ranger Kimmie Johnson patrolling one of Grand Canyon National Park's most heavily used trails, the Bright Angel. Zoroaster Temple looms in background.

The most numerous of the Grand Canyon's native peoples are the Athabaskan-speaking Navajo. Believed to have migrated south from Canada about a thousand years ago, "the first sure record of a Navajo visit to the Canyon is as late as 1863, when a group of them fled into the abyss to hide from Kit Carson" (Hughes, 1967, p. 13). Today, they inhabit a region that surrounds both the Hopi and the Little Colorado River Gorge Navajo Tribal Park. Known to themselves as the Dinéh, "the People"—"Navajo" was derived from the Spanish name Apaches de Nabajó—165,000 people now inhabit a 16-million-acre, coal-rich tract of the Colorado Plateau east of the Grand Canyon.

Between their cultural apex and its nexus, however, the Grand Canyon's native peoples watched, guided, and fought an erratic procession of missionaries, soldiers, trappers, explorers, prospectors, cowboys, surveyors, misfits, settlers, moonshiners, developers, and tourists who began swarming over the Grand Canyon not long after García López de Cárdenas first "discovered" it in 1540. Whereas these native peoples lived and breathed within its breach, revering the very source that sustained their spirit and life ways, many newcomers saw the canyon simply as a treasure chest ripe for the plucking.

Most storied among them, perhaps, was Maj. John Wesley Powell. He danced out the visions of Washington bureaucrats who saw the untapped waters of the Colorado River as the one sure way to open up and settle the parched West beyond the "dry line," the 100th meridian. They weren't far wrong. In the end, Powell's adventures through the "great unknown," perhaps more than the science he and his men gathered during their two unprecedented river surveys, advanced the "promotion of his own fame" (as Thomas F. Dawson wrote in 1917). On May 24, 1869, the one-armed explorer and a crew of eight men resumed their epic quest to run the Green and Colorado rivers all the way through the Grand Canyon. It was a quest filled with mutiny, revelation, and near disaster; a victorious Powell emerged from the Grand Canyon ninety-seven days after first leaving Green River, Wyoming. En route, three good men had abandoned his manic grip at "Separation Rapids," only to be mistakenly killed by Shivwits Paiutes during their trek out.

Nearly everyone believed Powell was "the first one through!" But he left a controversy in his wake that remains unsettled to this day. It concerns the simple, honest words of a sunburnt and emaciated prospector named James White. Near death, White drifted up to the banks of Callville, Nevada, on a crude log raft, claiming to have spent two weeks floating through the "Big Cañón" in a desperate attempt to escape hostile Ute Indians. If his incredible tale was true, the 30-year-old White was the first man in history to navigate the Colorado River through the Grand Canyon

OVERLEAF: *Dory boatman Renny Sumner rows through Conquistador Aisle near River Mile 121 in the Grand Canyon.*

Ancient Puebloan people who lived in the Grand Canyon stored food in rock structures such as this.

(albeit not for scientific reasons), two years before Powell's first successful voyage. Sides were taken in this controversy; battle lines were drawn. Detractors said it was impossible for a lone man to run a frenzied river choked with huge, boiling rapids, on a raft made of three cottonwood logs. Others cried foul, claiming Powell was less than honest for having written the popular account of his two Colorado River expeditions to sound like one adventure. Still others pointed out that Powell may well have already gotten wind of White's adventure from a newspaper article and thus realized there were no falls that couldn't be run by a well-equipped expedition. Still, it remains a riddle that will never be solved: Too many contradictions, too many half-truths and lies, too many players have long since been silenced by Father Time. It's a "Colorado Rivergate."

About the time White and Powell were becoming legends, prospectors began streaming toward the Grand Canyon for the untold riches they believed must surely exist in its dreamy abyss. Some like white-bearded hermit Louis Boucher, who was said to "Ride a white mule, and tell only

Fifty-seven miles long and over 3,000 feet deep, the Little Colorado River Gorge is one of the Grand Canyon's most spectacular tributary canyons.

white lies." Meanwhile, "Captain" John Hance discovered what most others eventually would: more money was to be made guiding dudes into the canyon than trying to drag low-grade ore out of it.

Yet because of such miners as Hance and Boucher, and others such as William Wallace Bass, whose ashes were scattered over the canyon after his death there in 1933, eighty-eight rim-to-river trails once penetrated the inner canyon. Many of them were handforged along ancient Indian routes. And visitors were all too eager to use them, especially the Bright Angel Trail; it dropped out of the burgeoning South Rim mecca, passed the oasis and former Havasupai encampment of Indian Gardens, and ended with a blissful night on the bottom of the Grand Canyon at storybook Phantom Ranch.

First trod by native peoples, rebuilt by prospectors to reach their sorry diggings, and later used by tourists in quest of the canyon's scenic wealth, the modern trails and ancient paths attracted a new kind of visitor to the Grand Canyon during the 1950s, and they weren't interested in gold any more than the canyon's original inhabitants were. A little hard-won glory, perhaps, but nothing more. They came to climb the canyon's majestic sandstone temples, soaring spires of rock, distant island mesas, and craggy buttes. Most of these hard-to-reach land-forms were named during the 1880s by eminent geologist Clarence Dutton after Eastern deities.

Foremost among Grand Canyon climbers was Harvey Butchart. The father of canyoneering if there ever was one, Butchart worked out the mind-numbing puzzle of the grand maze on foot, trekking over 15,000 miles, making over fifty first ascents, pioneering ninety-six rim-to-river routes, and finding 154 breaks through the relentless Redwall formation—the most intimidating barrier in the entire Grand Canyon. Like the canyoneers who soon tried to match his frenetic lifelong pace, however, Butchart and others discovered that many of the canyon's easier, though no less remote, temples were first climbed by native peoples—perhaps the Anasazi—as long as a thousand years ago. And only the most difficult "temples," such as Zoroaster, eluded the grasp of such veteran canyoneers and native peoples. That is, until Arizona climbers Dave Ganci and Rick Tidrick boldly stepped onto the scene. Applying rock-climbing techniques honed in Yosemite Valley, and a savviness for desert travel that eluded other parties, they climbed the great sandstone monolith of Zoroaster Temple on September 23, 1958, after days of struggling through mind-warping heat. In doing so, they snatched what, to climbers, has always been the Grand Canyon's biggest plum.

Nearly three decades after Ganci and Tidrick's audacious first ascent, a dark cloud had formed over the ancestral canyon lands from which its native people had been forcibly removed. As President Theodore Roosevelt said, "Leave it as it is. You cannot improve on it. The ages have been at work

The rarely seen ''other side'' of the heavily photographed Havasu Falls.

Pioneer Grand Canyon climber Dave Ganci rappels off the 6,761-foot summit of Angel's Gate.

at it, and man can only mar it" (Hughes, 1967, p. 102). No one listened, and the once pristine natural world of the Hopi, the Havasupai, the Walapai, the Southern Paiute, and the Dinéh came to an end. At the turn of the century, in a decade-long slaughter, Government hunter Uncle Jim Owens had killed 532 mountain lions on the North Rim. So the ecological balance between the North Rim's lion and deer population spun out of control, and thousands of deer died of starvation. The wolf's howl also fell silent, and the great Mexican jaguar fled back to Chihuahua and Sonora.

But that was only the beginning. Hotels, motels, gas stations, and curio shops began smothering ancient Anasazi encampments, proudly selling the fine jewelry of native people eagerly displaced by the white man. Ranger Rick changed, too; although he still wears his Smokey the Bear hat, now he's packing a .38. And just in case things get heavy, he's got a .12-gauge pump cradled near the front seat of his patrol car. That's because crime is now so rampant at the Grand Canyon, a U.S. magistrate works full time (officially "one-third time") to handle the tremendous caseload of felonies, misdemeanors, and traffic citations that flood his South Rim courtroom each week. As one ranger told me, "Everybody likes to visit a National Park—even the bad guys." Smog drifts in from the Navajo Power Plant, smothering the rimbound scenic vistas that 4 million camera- and video-toting tourists come by plane, train, and automobile to see each year. Even on the rarer, clear days, the skies over the canyon are so crowded with helicopters and airplanes that 227 people have perished in air crashes since two commercial airliners first collided over the eastern Grand Canyon in 1956. The staggering death toll has not stopped the 300,000 sightseers who continue to fly over the canyon each year. Nor do hikers welcome the relentless drone and "wup-wup-wup" of 100,000 helicopter and aircraft barnstorming the canyon each year. After all, they've paid stiff entrance fees to hike to the bottom of the greatest canyon on earth. But the trails are no less crowded than the skies; some 40,000 people hike and camp in the inner canyon each year. Most turn what would otherwise be a mesmerizing trek into a near-death experience because they come during peak summer vacation months, when the inner canyon desert is an inferno waiting to burn them alive. The outpost of Phantom Ranch looks like a war zone of tourists crippled by deep, skin-ripping blisters and heat exhaustion that damn near sucked the life out of them. Look at the river, the procession of rafts chugging, rowing, and paddling through the narrow corridor every summer day, and it's no less crowded than the trails, because 22,000 people stand in line each year to run the legendary "big drops," first braved by the likes of James White and Major Powell. They are forced to camp on the dwindling sands of tiny beaches that sometimes disappear before their very eyes, because upstream the Glen Canyon Dam traps an estimated 500,000 tons of sediment that once recharged the beaches each day. Because the

A coyote gasped his last breath on this mud-caked flat in the Little Colorado River Gorge.

river is now regulated for the "peak power" demands of distant cities like Phoenix, Las Vegas, and Los Angeles, modern river runners now brave man-made tides that have more in common with artificial wave machines like "Big Surf" than with the wild river that once cut the greatest canyon on earth.

Nobody listened to Roosevelt. And they certainly didn't heed the sage life ways of the native peoples who first dwelled here. Gotham was erected where once there were only clusters of small stone dwellings, and a sprawling, crime-ridden tourist mecca now hangs exactly where Roosevelt warned us not to build, on the very edge of the Big Cañón: "I hope you will not have a building of any kind, not a summer cottage, a hotel or anything else, to mar the wonderful grandeur, the sublimity, the great loveliness and beauty of the Canyon" (Hughes, 1967, p. 102). But it's done. And what is there to do about it now, really? Do we move it all back from the rim, as the late Edward Abbey first suggested? That's never going to fly here; because, in the case of the Grand Canyon, profit has always dictated common sense and destroyed what natural sense man might have had when he first thought he could "manage" an incomparable "resource" that never was, and never can be his.

Yes, if there truly is light after darkness, it is the view from the rim, the view from the river. Either way, you are peering through the opening through which mankind originally emerged. It is there in the sacred Hopi beliefs, for anyone who has the sense to listen. Because it is to this opening, the very womb of the earth, that each of us, every single one of us, is inextricably drawn. And that is the light after darkness, the sheer wonder of the natural force it continues to bear on our lives. The Grand Canyon of the Colorado River.

Cañón del Diablo
Baja California Norte

But Baja California . . . [is] a region of tumbled mountains,
yawning chasms, desert plains, lonely shores, barren islands.
It appears to be wholly unmarked by man.

WILLIAM WEBER JOHNSON
Baja California (1972)

For the most part, the desolate, 800-mile-long Baja Peninsula is still unmarked—especially the Sierra San Pedro Mártir: a modern no-man's land that claims the remote Observatorio Astronómico Nacional as its only alpine outpost.

Clawing at the heavens midway between the Pacific Ocean and the Sea of Cortez, the twin horns of 10,154-foot Picacho del Diablo, Peak of the Devil, crown the Sierra San Pedro Mártir—the highest and most rugged mountain range in Baja. Uniquely situated due south of the Colorado River Delta and El Gran Desierto, the largest sand sea in North America, the Sierra San Pedro Mártir, together with the 6,658-foot Sierra de Juárez, comprise a daunting 175-mile-long spine of raw granite that some have called the "backbone of Baja." Formed in the Cretaceous period 100 million years ago, geologists also tell us this border-hopping range is actually a continuation of southern California's peninsular range, which emanates on its northern end with the 10,800-foot San Jacinto Mountains and is physiographically linked to the Baja cordillera by California's Santa

75

Rosa and Laguna mountains. Consequently these mountains, along with El Parque Nacional de San Pedro Mártir and El Parque Nacional Constitución de 1857 (Sierra de Juárez) are currently being studied as a Transborder Biosphere Reserve. What further distinguishes Baja's loftiest sierra is that it is the only mountain range in North America hemmed in by two seas; so it's not a region that comes immediately to mind when you think of canyons.

Radiating from the summit crest of the Sierra San Pedro Mártir, however, is a labyrinth of dreadfully rugged *barrancas* (canyons), "Canyons and arroyos," that William Weber Johnson wrote (p. 22), were "carved by forgotten torrents, each ending in a dry alluvial fan on the desert floor." Named after both God and the devil, these barrancas have been variously called La Providencia (the Providence), Diablito (Little Devil), and Diablo. The most wicked of them all, Cañón del Diablo is the cordillera's greatest barranca, and it follows a natural faultline from the Mártir's soaring upheaval of granite into the subterranean sink of Laguna Diablo, Devil's Lake, 8,000 vertical feet below.

As early as 1928, Desert bighorn sheep hunter John Cudahy hit the mark when he described these barrancas as "a tortured region of chaotic immensity without a single feature of repose where all living things seem to have been exorcised by some spectral curse of banishment." Among those "exorcised" were native peoples like the Kiliwa, Akwa'ala, Ñakipa, and Juigrepa who survived in the shadows of these barrancas before the strange ways and exotic diseases of the "black robes," the Spanish missionaries, drove them over the edge of extinction. Of their life and spirit ways, little else is known about these vanished people other than they eked out a meager living hunting and gathering where no man can sustain himself today; and only *el león*, "the lion," still roams wild with their ancestral spirit through the hanging barrancas after the wary, sure-footed *borrego*, the Desert bighorn.

First seen by non-Indians in 1539 when Spanish explorer Francisco de Ulloa sailed up the Sea of Cortez, one of the earliest descriptions of the Sierra San Pedro Mártir was written on November 6, 1706. That's when the tireless Padre Eusebio Francisco Kino and his faithful entourage climbed Serro de Santa Clara for the second time. However, whereas Ulloa had unknowingly sailed into the cul-de-sac formed by the Colorado River Delta in hopes of reaching the West Indies, Padre Kino set his sights on finding a land route to Pimería Alta, "upper Pima land," to Alta California. From the 4,235-foot summit of Cerro de Santa Clara (known by Anglos as Pinacate Peak) in the frontier of Sonora, the missionary explorers again surveyed the shimmering sweep of sand and sea to the west; in so doing,

A canyoneer leaps across swollen creek waters near the mouth of Cañón del Diablo in the Sierra San Pedro Mártir of Baja California Norte.

Iron pyrite, ''fool's gold,'' glittering in this small cascade, once lured prospectors into Cañón del Diablo.

they glimpsed the Sierra San Pedro Mártir 125 miles distant and realized that Baja California, or Lower California, was not an island as they'd originally believed. They called the towering cordillera California's Sierra Madre; wrote expedition diarist Alférez Juan Matheo Ramires, "And we saw that the Sierra Madre of California runs from south to north to where the sea ends, and that a point shuts in a bay which Fray Manuel calls the estuary, because it is the mouth of the Río Colorado, at the head of the Sea of California."

Since that historic *entrada,* "entrance," or journey, Kino's Sierra Madre has been mistakenly called both the Calahamue Mountains after Arroyo Calagnujuet, one hundred miles south of Baja California Norte's 10,154-foot-high point, and the Santa Catalina Mountains after the mission by that name in the Sierra de Juárez fifty miles north. It wasn't until Misión San Pedro Mártir de Verona was founded on the western flanks of the Mártir in 1794 that the range had an accurate namesake; and Sierra San Pedro Mártir is now the official name that appears on the error-plagued topographical maps of the region.

Few early explorers or climbers had accurate maps to guide them out of the burning white salt pan of the lower Sonoran desert at the eastern foot of the range through the four distinct life zones that lead up to the Canadian forest that shrouds the Mártir's highest reaches. Described by historians as "among the great western pathfinders," Father Wenceslaus Linck made the first recorded traverse of the Sierra San Pedro Mártir, using a canyon system to forge a route down the impregnable east face of the range in March 1766. But it wasn't until 1911 that the mountain received its first recorded ascent.

Even climbers—normally united by their common love of mountains—have come to view this bewitching mountain with two sets of eyes. Americans—lured by the irresistible challenge and tales of climbing what writer and Baja historian John W. Robinson accurately described as "one of the finest mountains on the North American Continent"—call it Picacho del Diablo, Peak of the Devil. Mexican *alpinistas*—who see the heavens where gringos still see the devil—call the twin-summited peak by one of its two earliest names, Cerro de la Encantada, Mountain of the Enchanted One, or La Providencia, the Providence.

By any name, climbing journals and popular accounts are rife with the tragic exploits of those who met the mountain and gave up the ghost—or lost. But the first known person to call it the Providence after climbing it solo was Donald McClain; the California topographer came from the era of "wooden ships and iron men" but, even on this storied mountain, his was not an ordinary tale.

Gazing northeast from the North Summit, you can just about trace McClain's epic March 1911 journey: first by skiff from Yuma, Arizona, down the silver ribbon of the lower Colorado River slithering through the fertile delta to the Sea of Cortez; then by foot across the mirage-shrouded wastes of the San Felipe Desert. But once McClain reached the Gulf port of San Felipe, he trekked west across the Desierto en Medio, "Middle Desert," to the foot of the cordillera and traversed the crest of the range by following an ancient Indian route up Cañón del Cajón, a half-dozen canyons north of Father Wenceslaus Linck's route. But McClain hadn't struggled all the way

OVERLEAF: *The serrated summit crest of the Sierra San Pedro Mártir.*

Summit register for Picacho del Diablo.

from Yuma to climb the mountain he preferred to call La Providencia; he was looking for placer mines. That is, he was until he stood on the yawning vista atop the imposing headwall of Cañón del Diablo and saw the Providence again. Then and there, he decided to climb the mountain on sight.

Carrying little more than a wheel gun, a pouch of jerked meat, a blanket, and a jacket, McClain traced the careful words of a Mexican miner into the depths of Cañón del Diablo and turned out of it along another *minero*'s route that led to a gloryhole high on the mountain's northwest

Climber Jason Lohman stands atop the 10,154-foot summit of Picacho del Diablo, the highest point in Baja California.

flank. Driven by supernal forces, McClain reached this alluring perch and when he did he wrote, "with the help of God who 'softly and tenderly called me today' led me up the north ridge of this noble mountain to the summit and beyond."

Beyond? McClain may have been alluding to his daunting return journey; because after reaching the summit, he was later quoted as saying, "There was no evidence of any previous visit. I found it hard to believe that no one had been there before. . . . I did not take time to build a cairn because of the uncertainty as to how I was going to retrace my route."

He did, and in the years since others followed McClain's lead—maybe a thousand all told. Some came from the west across the forested table-lands, as the 1932 Sierra Club party did when they made the second ascent of the mountain. Still others came from the east, up the boulder-choked depths of Cañón del Diablo, as I did. But none matched the three-hundred-mile summit journey of Donald McClain. Not even those who left their bones in what they knew as Canyon of the Devil.

Cañones de la Isla Tiburón

Sonora

> The very air here is miraculous, and the outlines of reality change
> with the moment. The sky sucks up the land and disgorges it.
> A dream hangs over the whole region, a brooding kind of hallucination.
>
> JOHN STEINBECK
> *The Log from the Sea of Cortez* (1941)

Perhaps no other hallucination has hung over this region longer than the mythical apparition of Isla Tiburón, Shark Island. It shimmers in the pale, moonlit sea west of Bahía Kino, Sonora, like a great leatherback turtle breaching for air. To scientists, *Dermochelys coriacea* is the largest sea turtle in the world; to the Seri Indians, who both revered and hunted their godlike creature, it is *moosnípol,* "sea turtle's blackness."

It's not without some irony, then, that the ancient topography of the Seri's vanquished ancestral lands bears a striking resemblance to the carapace of this nearly extinct reptile. Twenty miles wide and up to thirty-five miles long, the 750-square-mile Isla Tiburón is split lengthwise by two seldom-visited mountain ranges; both fall within the Basin and Range physiographic province. The Sierra Menor runs along the west side of this Sonoran desert island and forms a rugged coast that mirrors the setting sun, while to the east the Sierra Kunkaak erupts out of the *bajadas* (lowlands) of the Valle de Tecomote. Of the two, the 2,871-foot Sierra Kunkaak, "Mountains of the Seri People," is the highest, and its steep, east-facing barrancas the most storied.

Like Baja California Norte's Sierra San Pedro Mártir two hundred miles to the northwest, Isla Tiburón was also "discovered" by Spaniard Francisco de Ulloa in 1539 while he was sailing north through the Sea of Cortez to the West Indies—or so he thought. Reading the ancient eelgrass, however, archaeologists tell us the Seri Indians inhabited Isla Tiburón for at least 2,000 years. Some believe that was about the time the original Seri—the ancient Giants known by them as *xica coosyatoj,* or "things singers"—migrated from the east coast of the Baja Peninsula. Using the Midriff Islands of San Lorenzo, San Esteban, and Tiburón as stepping-stones, the Seri navigated the treacherous Gulf currents on fragile *balsas* (rafts made from bundles of reedgrass or cane) and subsequently dispersed into six regional bands that roamed Isla San Esteban, Isla Tiburón, and the west coast of Sonora in a desperate, lifelong quest for food and water. Until their population collapse near the turn of the century, however, most Seri lived on Tiburón, because the island's diverse flora and fauna offered these tenacious people the best chances of surviving the arid reaches between the desert and the sea.

But that was before 1965, the year the Mexican government turned Isla Tiburón into an ecological reserve and made way for the twenty Desert bighorn sheep that were transplanted there in place of the Seri ten years later. Consequently, those Seri who hadn't already moved to the mainland to pursue commercial fishing opportunities supported by the Mexican government were forced to relocate and settle in one of two principal villages: Punta Chueca, 40 kilometers north of modern Kino Bay; and El Desemboque, an ancient Seri encampment once known as Haxöl Ihoom, "the place of the clams," another 75 hard klicks by sandy track north of Punta Chueca.

Separated from the central Gulf Coast of Sonora by the Canal del Infiernillo, Little Channel of Hell, Isla Tiburón is clearly visible to the estimated 550 Seri now wresting a living from those arid coastal villages weaving baskets, making ironwood carvings, and fishing Gulf waters nearly depleted by factory fishing boats. Yet for most Tiburón still remains a physical and spiritual landmark that forms the center of the Seri universe; it is the home of the god Hant Hasóoma, "land *ramada,*" who first created the Kunkaak, the Seri people, on Tiburón. To the handful of Mexican Marines occupying the island's only dwellings, however, it is little more than a hardship post. In any case, Shark Island remains a wild, untracked land, where mystery and legends never died, where *cok-sar*—outsiders—sometimes did.

The unwary and the adventuresome have always been lured and taunted by the mysterious Cañones del Tiburón. Riding the twilight winds across the Canal del Infiernillo in a *panga,* a small motorized boat, piloted by Seri Ernesto Molino, the dark reflection of Isla Tiburón evokes the

Seri boys play, mid-afternoon near Desemboque, Sonora.

A collared lizard basks in the warm sun on Isla Tiburón.

haunting image of what many early explorers believed: that to visit this no-man's land was akin to visiting a world that time forgot. Still, they came.

Some who went never returned. Fact or fancy, the lore started accumulating in March 1700 when Juan Bautista de Escalante raided the *tiburones,* the Tiburón band of Seri. It's not known how many people were killed in that vicious clash, but in May 1894 gringo adventurers led by San Francisco reporter and pioneer Grand Canyon boatman George Flavell and another "newspaper man named Robinson," were reportedly killed by Seri while exploring the north side of Isla Tiburón. Two years later, prospectors George Porter and John Johnson were also killed on Tiburón. No one seems to know who or what provoked those killings, but the death of four Americans at the hands of "wild" Seri Indians prompted many Americans to believe the Seri were cannibals.

Those unsubstantiated claims didn't stop the four-man Grindell expedition from trying to reach Tiburón during June 1905 in hopes of discovering its rumored trove of gold. But the four Americans never even reached the island. In fact, near the modern Seri village of El Desemboque they found the withered, sun-jerked hands of two white men nailed to a wooden plank—hands one historian later believed were those of two Los Angeles prospectors named Olander and Miller, who'd sailed over two hundred miles south from Yuma down the Lower Colorado River and across the Sea of Cortez, via Isla Tiburón.

But that ghastly milepost only foreshadowed the tragedy that was about to befall the ill-fated Grindell expedition. Three men perished of thirst in the burning Desierto del Purgatorio—Desert Purgatory—even before setting sail for Tiburón and the fourth, John Hoffman, only managed to survive by making an extraordinary, often delirious, survival trek to the Gulf Port of Guaymas.

In light of its forbidding reputation among *cok-sar,* it's easy to understand why Isla Tiburón remained virtually unexplored by non-Seri until December 1921; that's when hunter and naturalist Charles Sheldon—against the advice of terror-struck local Mexicans—journeyed to Tiburón, alone, in the company of Seri guides to hunt *buro* (deer) for their starving families.

That Sheldon did, much to the gratitude of a small band of famished Seri still living on Tiburón. But the Yale graduate's otherwise descriptive journals provide few clues as to his exact course of travel in Tiburón's rugged canyons. And, strangely, of the 1.3 million square miles of Mexico mapped by the Dirección General de Geografía del Territorio Nacional, only Isla Tiburón remains uncharted today. In short, when medieval cartographers pointed to the blank spots on their map and said, "Here be dragons," they might as well have been talking about Isla Tiburón. Except for a *Gemini V* photograph taken from 100 miles up, Isla Tiburón remains *terra incognita* for most *cok-sar* today.

For most Seri, however, Isla Tiburón offered them the best chance of surviving in a stark region of the Gulf Coastal phase of the Sonoran desert, a virtually waterless region that claimed a meager Pleistocene carrying capacity of 0.9 people per square mile: the Xica Hast Ano Coii, "they who live in the mountains," inhabited both the neighboring Isla San Esteban and the southwest coast of Tiburón, reportedly "one of the hottest and most

arid environments ever to be permanently occupied by humans"; the Tahéöjc Comcáac band, or Tiburón-Island people, inhabited the remaining coastal zones of the island and the adjacent coast of Sonora; while the Heno Comcáac, "desert people," inhabited the rugged canyon regions of Tiburón. Each of these bands were masters of an austere coastal desert bioregion that offered the canny hunter and gatherer a diverse variety of food sources: from the once abundant *cahuama* (Pacific green turtle) to the brown pelican, mule deer, chuckawalla (large lizard), cactus fruit such as *pitaya*, and grasshoppers—among the many other food sources the Seri consumed and sometimes revered.

Like desert people everywhere, the Seri were also linked by a tireless quest for water. But of the forty-three water sources on Tiburón reportedly named by the Seri, "Only twelve or thirteen of these were considered permanent." Most were *tinajas,* rain pockets located in the *cañones* of the Sierra Kunkaak. Suffice it to say, the Seri's hardscrabble existence depended on those precious *tinajas.* But for those living in sere distant beach encampments, grueling treks were required to reach these canyon *tinajas* before their life-sustaining waters could be carried back in fragile, eggshell-thin clay pots. Suspended in straw nets carried on a yoke over one shoulder, these ollas were sometimes dropped by thirsty Seri and today it's still common to see Tiburón's ancient trails littered with shards of broken pots, what archaeologists call Tiburón Island Thinware and what locals call *cáscaras de huevos,* "eggshells."

Of Tiburón's ancient canyon trails, perhaps none was more heavily traveled by the Seri than the incipient path that can still be followed from the prehistoric beach encampment of Zozni Cmiipla, through the bajada's bewitching clusters of ocotillo, *cardón* cacti, and sacred *palo blanco* trees to the canyon-rimmed Pazj Hax, "graphite water." It is the only living water on Tiburón and, next to Xapij, "reedgrass water," it is only one of two principal locations the Seri gathered—or grew, as some speculate—reedgrass and giant cane to make their *hascám,* balsa rafts. First seen by Padre Adamo Gilg in 1692, Juan Bautista de Escalante also described the balsa in March 1700: "The Seri cross [the Infiernillo] in balsas composed of many slender reeds, disposed in bundles, thick in the middle and narrow at the ends. . . . These balsas sustain the weight of four or five persons and cut the water easily" (Felger and Moser, 1985, pp. 310–311).

Looking east from the 2,871-foot summit of the Sierra Kunkaak, the view encompasses a rugged unnamed cañón once hunted by the Seri Indians, the bajadas *(lowlands) fanning out to the coast, and the* Canal del Infiernillo, *"Little Channel of Hell," which separates Isla Tiburón from the west coast of Sonora, Mexico.*

OVERLEAF: *Seri boys play with a toy boat, while a* panga, *"motorized boat," heads into the* Canal de Infiernillo *for a night of fishing near Desemboque, Sonora.*

The Molino family sits for a portrait at home in Desemboque, Sonora. Nora (standing, left) and Aurelia (standing, right) are gifted Seri basket makers.

The first known *cok-sar* to reach Pazj Hax was ethnographer William J. McGee and a thirst-ravaged crew that included one Yaqui Indian, five Papago Indians, four gringos, and two Mexicanos "armed to the teeth." After reaching the ancient waterhole on December 19, 1895, they named it Tinaja Anita after their skiff, the *Anita.* Of their grueling search for water, McGee wrote,

> [We] set out to hunt for water. Toward the "yellow spot," past it we plodded, and on up the canyon following a fresh Seri trail . . . and on until darkness

A palo blanco tree (left) and sotol grow together among lichen-covered rocks near the summit crest of the Sierra Kunkaak, Isla Tiburón.

began to fall and Don Ignacio began to be discouraged; at last, looking up the deep barranca from a little knob, I saw it and against an abrupt mountain wall in a narrow gateway; and just below the narrow gateway I made out in the dusk a most welcome spot of verdure. On we went and at last reached a little patch of canes (a *carrizalito*) with water slipping over the rocks in nooks and potholes. How we drank. (Carmony and Brown, 1983, p. 39)

What proved to be the life-saving waters of Tinaja Anita, also proved to be a pivotal stop for *Kon Tiki* author and adventurer Thor Heyerdahl

Seri basket weaver Rosa Barnett uses a dagger-sharp deer bone awl to weave the dyed, split-bark of the torote bush through the coils of a sapim, *a huge basket that, when completed, will be sold to a collector or museum.*

three-quarters of a century later, when he first researched the Seri's use of the reed balsa for his famous Ra Expeditions. In the book by that name, Thor Heyerdahl (1971, p. 54) wrote

> This time [the Seri guide] pointed with his whole hand. . . . High up on the opposite mountainside the bare red rock split in a branch canyon running from a little plateau, and up there the sun was shining on a lush green patch, more fertile and lush in its light spring-green colour than any cactus or desert plant. Reeds! . . . The Seri Indians had not built reed boats because they had easy access to reeds. On the contrary, they had made their way right up into the mountains [Sierra Kunkaak] to find a minute trickle of fresh water where they could plant reeds to provide the raw materials for their boats.

Heyerdahl traveled around the world to explore this remote island canyon in hopes that the knowledge he gleaned would help him sail 4,000 nautical miles across the Atlantic Ocean from Morocco to Barbados. It did. And looking east from the summit of the Sierra Kunkaak, down its immense headwall into the miraculous light of the dawn seas, it's easy to see how the Seri linked their secret canyon to the sea. Listening to the voice of Hant Caai, the Seri god first told the *kunkaak* to build their seafaring balsas, and today you can't help but wonder what it must have been like for those first ancient Seri to ply the Canal del Infiernillo in balsas made from these canyon reeds, hunting the great black shadow of *moosnípal*, that swam where the mouth of this lost canyon empties into the sea.

The Canyons of
Big Bend
Texas and Coahuila

The Big Bend region . . . is one of the great wonders of the
Southwest. Wild, vast, and isolated, this land of desert,
river canyon, and rugged mountains is as close to the primeval
as anything on this continent.

RONNIE C. TYLER
The Big Bend (1975)

Tyler was not wrong. Seen through the shimmering purple veil of twilight, the only form that can be made out from the summit crest of Mexico's Sierra del Carmen is a ribbon of quicksilver slithering across the dark and empty ground five thousand vertical feet below. People living to the north tell us it's the Río Grande, the Big River. But the people living to the south have always said it was the Río Bravo del Norte, Brave River of the North. People toiling on opposite banks of this languid river, by either name, are told to believe that nothing exists on the other side. Look at any highway map of Texas, and south of this river the world does not exist; it is a blank spot. Or look at any highway map of the Mexican states of Coahuila or Chihuahua; north of this same river, the world falls away. The view from the region's loftiest mountain confirms this: nothing exists on either side of this river, except an ocean of desert, riled by angry mountains, tormented by riverine gorges. Come daybreak, the story will not change; it hasn't for four centuries. That's when Spaniards first proclaimed this fearsome ground a *despoblado,* "uninhabited land." And it remains just that: the largest, most desolate region on the entire 2,000-mile-long U.S.–Mexico border. Only, today this wild country is called the Big Bend for the dogleg turn the river makes, and the story of its canyons has always been told where the river runs through it.

Born from the cordilleran snowfields of southwestern Colorado, the Río Grande tracks south out of the Rocky Mountains and snakes its way across the New Mexican desert through the Jornada del Muerto—the Journey of the Deadman—depression before slipping through the border cities of El Paso del Norte, Texas, and Ciudad Juarez, Chihuahua. Here, where it has always been called the Río Bravo del Norte, the river turns southeast and forms the natural border between Chihuahua and Texas, as it contours around the northeastern foothills of the Sierra Madre Occidental. Were it not for the Río Conchos, the Río Bravo would have all but disappeared by the time it reached Ojinaga, but after trundling off the crest of the western Sierra Madre the Río Conchos breathes new life into it. Here, at what was once called La Junta, The Junction, the Río Bravo picks up the force it has always needed to carve the canyons of Big Bend and to find its way to the Gulf of Mexico, 1,896 miles from its source.

Cupped between Mexico's two greatest mountain chains, the Chihuahuan Desert fills the basin between the Sierra Madre Occidental and the Sierra Madre Oriental. Where the Río Bravo hunts its way across the northern reaches of this hellish desert to the eastern Sierra Madre, Big Bend's three most spectacular river canyons are found. Created some 3 million years ago, the 1,516-foot-deep Santa Elena Canyon was formed when the river sluiced a 19-mile-long gorge through the 3,883-foot Mesa de Anguila and Chihuahua's Sierra de Ponce. Nearly fifty river miles downstream, the 1,700-foot-deep Mariscal Canyon was formed when the river made its namesake, 6-mile-long bend through 3,932-foot Mariscal Mountain and Coahuila's Sierra de San Vincente. To carve the region's greatest canyon another 26 miles downstream, the river had to gouge a serpentine, 30-mile-long chasm through the one mountain range that has always linked the Rocky Mountains with the Sierra Madre del Oriental: the 8,400-foot Sierra del Carmen. North of the line, this border-hopping range takes the adopted name of Sierra del Caballo Muerto, the Dead Horse Mountains, for a party of U.S. surveyors who pole-axed nine of their horses they feared would be stolen by Indians in 1881. South of the river, the limestone-cliff-terraced sierra retains its birthright and serves as a lost refuge for the diverse array of indigenous plants and animals that have always shared the bioregion of the two countries.

What was first viewed as a border not long after the Conchos marched into extinction once linked the two dispersed bands of these Uto-Aztecan-speaking Indians. Ethnographers believe the Conchos were either buffalo

Twilight along the Río Bravo del Norte.

hunters, or possibly descendants of the northern Pueblo Indians who migrated down the Río Grande, before being subjugated by Spaniards during the seventeenth century. Until they were, the sedentary Jumano band lived near La Junta and survived largely on the labors of a digging stick economy. The nomadic Chisos band survived by their wits in the lofty reaches of Big Bend's 7,825-foot Chisos Mountains and the nightmarish deserts and canyons that fanned out below, hunting virtually anything that moved: from razor-tusked javelina and timid turtles to deer that the fleet-footed Chisos ran down, much as the Tarahumara did in the sierras and barrancas far to the southwest. Perhaps even more innovative than the canteens the Chisos made from horse intestines was their method for hunting ducks; canny swimmers, wearing masks fashioned from gourds, would float up to an unsuspecting bird and pull it beneath the glassine surface of the river before it could utter Quack One. Little else is known of the Chisos other than that they were vanquished by Spaniards in 1693 and corralled with the Jumanos until the last surviving members stared into the cold unblinking eyes of cultural extinction.

Agave, like these growing near the summit of the Sierra del Carmen, once provided the pulp for sotol, *a volatile tequila-like liquor smuggled across the Río Bravo del Norte during the U.S. Prohibition period.*

What was home to the Conchos was greatly feared by the Spaniards during their sixteenth-century *entradas* throughout the northern reaches of New Spain. Believing the area to be a deadly no-man's land, devoid of plant and animal life, and inhabited solely by murderous Indians, the Spaniards avoided the Big Bend region like the plague that shadowed them and ravaged the region's indigenous people. Not until the Pedro de Rábago y

Terán expedition crossed the Sierra del Carmen and passed within the breach of the Chisos Mountains in 1747 did the first non-Indians set foot in Big Bend.

Rábago y Terán was one of the lucky ones. Nearly every other party who ventured into the Big Bend throughout the century-long reign of terror that gripped the region came under siege by Mescalero Apaches or Comanches. Driven south from their own ancestral lands by U.S. settlers and the troops who guarded them on their quest for Manifest Destiny, both tribes fled south and quickly filled the ecological void left by the Chisos. But even as masters of this cruel, new environment, life in the desperate haunts of Big Bend was never promising for the Mescaleros or Comanches. Continually faced with imminent starvation, or the booty taunting them from nearby Mexico, the natural selection for both tribes was soon written in a trail of blood. They stole wherever they could, often from immigrants and freight wagons struggling along the dust-choked Chihuahua Trail between San Antonio and Ciudad Chihuahua. Expert horsemen and guerrilla fighters, the Mescaleros and Comanches soon began making lightning strikes against terror-stricken settlements on both sides of the river. Of the two, the Comanches embarked on the most far-flung raids, riding deep into the heartland of Mexico, as far south as Durango and Zacatecas. Their forbidding reputation grew as a result of these bloody raids, and people from those interior provinces soon came to fear nothing more than September's Comanche Moon. In this month the largest bands of horseback Comanches would ford the river between Santa Elena and Mariscal canyons to conduct their audacious raids a thousand miles south. The path they left in their wake was so heavily beaten—littered with the bleached bones of stolen stock, and the hideously bloated corpses of kidnapped women and children—that the Comanche War Trail became the official border between Chihuahua and Coahuila!

Mexico was so incensed by the Comanches' vicious plundering that the governments of both Chihuahua and Coahuila hired scalp hunters to work the canyons of Big Bend. None was more feared than the sinister blue-eyed gringo called Santiago. At $200 a scalp, the first of many illegal businesses flourished in Big Bend. As Ronnie Tyler wrote (1975, p. 70) in his enlightening, wonderfully lurid, history of Big Bend, scalp hunters "were not always too careful about whose scalp they collected, because it was difficult to distinguish among Anglos, Mexicans, or Indians once the lock had dried and shriveled."

While heads were rolling in Big Bend, candelilla soon found a market throughout the United States. Wax processed from the native plant (*Euphorbia antisyphilitica*) could be used in products ranging from candles to chewing gum. Harvested by poor *campesinos* on both sides of the river for $2.50 a ton, burro-loads of contraband wax were often smuggled out of

*Juan Valdez stands in front of his one-room adobe in the pueblo of
Boquillas del Carmen, Coahuila.*

Mexico's secret canyons once official Mexican government quotas were
met. Message senders, *avisadores,* used hand-held mirrors to signal look-
outs when all was clear and loads of precious wax could be packed across
the river. During Prohibition in the United States, moonshiners and
bootleggers used those same routes to smuggle *sotol,* a tequila-like liquor
craved by ranchers, cowboys, miners, dirt farmers, and settlers living on the
U.S. side of the river.

The dust kicked up along Big Bend's smuggling routes never settled
for long, not even after 707,894 acres of Big Bend was decreed a National
Park in 1944. Where quart bottles and five-gallon wooden casks of volatile
sotol once flowed out of Mexico's hidden canyons, bales of *mota* (marijuana leaves) and shrink-wrapped kilos of *sensemilla* (the potent marijuana
flowers) were floated across the river in *chalupas* (rowboats) to meet the
heady rush of the 1960s and 1970s. However, no contraband has choked
Big Bend's canyons more than *cocaína.* As cocaine was smuggled into
Mexico from Columbia and Peru by plane and boat, Ojinaga became the

*Little more than this small cemetery remains of Terlingua Abaja, a Mexican village
that once struggled for a toehold in Texas' harsh Big Bend frontier.*

principal staging area for a succession of ruthless drug cartels who strug-
gled to meet the gringo's insatiable appetite for the white powder. Of Las
Movidas, the drug underworlds, none was more infamous than Pablo
Acosta's. Few Chihuahuenses knew the border and its canyons better than
Pablo Acosta, who was born in the dirt-poor village of Nueva Santa Helena,
Mexico, across the river from the United States' most desolate National
Park near Santa Elena Canyon.

Viewed in Ojinaga both as Robin Hood and as a murdering drug dealer,
Pablo Acosta controlled his *plaza,* or drug concession, with the help of en-
forcers such as El Carnicero de Ojinaga, the Butcher of Ojinaga, who used
torture, mayhem, and death against Acosta's known enemies and suspected
traitors. For those Ojinagans who avoided the *chicharra,* the electric cattle
prod, and knew only Pablo's generosity, he was a hero. On April 24, 1987,
however, El Padrino, the Godfather of Ojinaga, went down in firestorms
of gunfire not far from the adobe *jacal,* or hut, where he was born on
January 26, 1937.

Ricardo Moran Espinoza waits for the afternoon surge of American tourists to belly-up to the bar in Boquillas del Carmen, Coahuila.

With *Drug Lord: The Life and Death of a Mexican Kingpin*, Terrence Poppa's riveting and daring exposé of Big Bend's notorious drug trade, it's easy to reconstruct the final moments of Pablo Acosta's life when viewing the burned-out ruins of his adobe in Santa Helena today. Frustrated with their inability to nail Acosta in their own country, the Mexican Federal Judicial Police—with the help of the FBI, Border Patrol, and National Park Service—staged a twilight Comanche-style raid from the Big Bend National Park side of the river. The plan worked. Two Bell 212 helicopters, carrying Policía Judicial Federal armed with U.S. AR-15's and Russian Kalasnikovs, literally dropped out of the sky onto the rooftops of Pablo's small pueblo. Never suspecting they would be attacked from the rear, (the U.S. side of the river), Acosta's men were taken by complete surprise; three dozen *pistoleros* (gunmen) and *gatilleros* (hit men) threw down their automatic weapons while the *padrino* fortified himself in his small adobe. But there was no way out, and Pablo vowed never to be taken alive. He died the way he lived.

Those who still sing of this infamous *contrabandista,* or smuggler, say he still lives. They sing that Pablo was smart; that everybody was on the take, up and down the line. That the military was bought, and the raid was staged to throw his enemies off his track. *¿Quién vive?* "Who lives?" *¡Vive Pablo!* "Pablo lives!" You need only ask around the *pueblitos* that line the river across from Big Bend National Park, such as Santa Helena or Boquillas, and the former Spanish presidios of San Vincente or San Carlos; ask *la gente,* "the people," what they thought about Pablo Acosta. Those who don't turn away, or flash you with a pearl-handled .45 automatic, are likely to tell you he still lives. Most will point north. *Las drogas,* the drugs, are not their problem; it's *"los pinches gringos."* Mexicans have many examples. The one you hear about most on both sides of the river is about the sheriff in Marfa, Texas, busted just north of the park with a ton of raw coke while the U.S. Customs drug surveillance balloon flew overhead. No, the smuggling through the canyons of Big Bend National Park will continue as long as *los gringos ricos*—the rich Americans—want to pack their noses or torch their futures in a glass pipe. And poverty's children will eagerly wear the boots of men like Pablo Acosta. It's there, all along the border, singing through the soul of the frontier's *corridos,* or ballads.

> Gone is Pablito, friend of the poor,
> Killed by the government
> In a world that shows no mercy
> For people like that.
> And the gringos, laughing on the other side of the river,
> Prayed for Pablito to die.
> Yet he had done nothing more
> Than give them what they wanted.

Nowhere else in the National Park system is a bioregion so awash in narcodollars and so divided by the socioeconomic and political agendas of two countries. Yet perhaps nowhere else in the world do two countries have the opportunity to create one of the largest protected bioregions on earth. First studied as an international park in 1935, *Parque Nacional de la Gran Comba* fell through the cracks for over half a century. Not until Eliseo Mendoza Berrueto, the governor of Coahuila, resumed discussions with National Park officials in 1988 did enthusiasm for an international park rekindle on both sides of the border. If established along the lines of Montana and Alberta's Glacier/Waterton International Peace Park, the combined 1 million square acres of Big Bend National Park, Big Bend Ranch State Park, and Black Gap Wildlife Management Area, along with the proposed 1.2 million square acres of Coahuila's Sierra del Carmen environs, would "create one of the largest protected ecosystems in the world."

But the linchpin for the "Tex-Mex Park" is the Sierra del Carmen. Long viewed as an outlier of Texas's mighty Chisos Mountains, the Chisos are little more than a hummock compared to the Sierra del Carmen and, further to the south, the Sierra Fronteriza. Yet the one can't seem to fully live without the other. Black bear, extinct for decades in Big Bend, have been migrating from the Sierra del Carmen and re-establishing in the Chisos. The Sierra del Carmen Whitetail deer, seen throughout the Chisos today, is also believed to have migrated from the Sierra del Carmen long before the two mountain islands were separated by the vast sweep of Chihuahuan Desert. *La pantera,* or panther, completes the natural cycle by feeding on the Carmen Whitetail, and now mountain lions are seen with startling regularity by hikers probing the timbered upper reaches of the Chisos. Plants, especially, ignore the political boundaries of this bioregion. Take the Chisos agave; once thought to grow only in the Chisos, it's now also found in the Sierra del Carmen. Closer to the river's edge, *bosques* (groves) of honey mesquite crowd both the U.S. and Mexican shores. In fact, climb up from the river in either direction—toward the Chisos or, better yet, the Sierra del Carmen—and you will climb through the same plant communities that fan out from the river in both directions. The desert scrub community, for instance, claims no less than 70 different species of cacti, among them the ocotillo, "the devil's walking stick." Clusters of *lechuguilla,* "shin daggers," and century plants fence each other in the wind as you climb higher through the grassland community carpeting both sides of the river. Only in the loftier reaches of the Sierra del Carmen do you find the virgin mix of grama grasses that escaped Big Bend's voracious cattle herds. And still even higher in the Chisos and Sierra del Carmen, you find Mexican *piñón* pine, alligator juniper, and western yellow pine, biologically linking both summits 50 air miles apart. In spite of their similarities, though, there's no comparing the stature of the two ranges. In *Naturalist's Mexico,* author Roland H. Wauer estimated that only ten square miles of the Chisos lie above 5,500 feet; whereas the Sierra del Carmen has over a hundred square miles that soar above that elevation contour. As a result, Wauer wrote, the Sierra del Carmen has a "greater variety of breeding raptors. . . . Golden eagles, redtails, zone-tails, goshawks, sharp-shinned and Cooper's hawks, peregrines and prairie falcons."

In short, the region begs to become an International Park were it not for one glaring fact: the United States is still struggling to come to terms with the one country it has always depended on. There is no better example of this dependency than in the small pueblo of Boquillas del Carmen. Located across the river from Rio Grande Village, Boquillas was founded at the mouth of La Cañón del Boquillas before the turn of the century. Lead and zinc ore from the Sierra del Carmen's Corte Madera Mine was transported across the river by a 6-mile-long aerial tram. And for a short time

A wall of Rocknettles reaches out for the fickle winter sun.

Boquillas actually prospered, largely by the toil and sweat of its *mineros* (miners) while the Americans grew wealthy from the ore. But today Boquillas remains a poor village of some two hundred God-fearing people who are struggling to gain a toehold in the twentieth century—*without* succumbing to the temptations proffered to them by the region's *narcotraficantes*. But they have been told they must continue to live in the past.

Here's why. Promised by the National Park Service that an electrical hookup would be strung across the river much as the park had done for the

*Silvia Ureste and her children in the one-room dwelling they share in
Boquillas del Carmen, Coahuila.*

neighboring river pueblos of San Vincente and Santa Helena, the people of
Boquillas scraped together $100 apiece—what amounts to three months'
wages on the frontier—to put electrical hookups in their small adobe
dwellings. Several power lines went up. Hopes soared. Two weeks, they
were told, and they too would have the luxury of dangling a bare lightbulb
over their kitchen tables.

But just when they were about to flip on the switch, the Park was sued
by two of America's most powerful conservation groups. The line, the
people of Boquillas were told, would endanger the peregrine falcon—
although the industrial tramway across the river was never a problem
when the Americans wanted the Sierra del Carmen ore. Besides, bring
these good people out of the dark ages, and the next thing you know
Boquillas will become another border town like Ojinaga. Nobody stopped
to think that the only way you can get to Boquillas from the north is by
rowboat. And from the south, it's even more difficult. Boquillas, you see, is
tucked away on the northernmost reaches of the Coahuilan frontier, and it

John Robson on a canoe portage at the Río Grande during a weeklong Executive Outward Bound course through Big Bend National Park, Texas.

takes three hours of hard driving on a rough dirt track to reach it from Musquiz, the nearest outpost. So while Americans hunker down across the river in $100,000 RVs, equipped with satellite dishes, and towing $20,000 Pathfinders, the people of Boquillas are told they must continue to burn lamp oil and butane while they watch the lights from Rio Grande Village blaze throughout the long, cold winter nights. To live out this romantic vision for the norteamericanos, though, they must ferry empty five-gallon jerry cans across the river in *chalupas,* and fill them with expensive gas from the Park store, so they can fuel their pickups to make the hard run into the nearest Coahuilan outpost that sells lamp oil and butane. Ask them if *they* think living in the eighteenth century is as quaint as U.S. tourists think it is every time the tourists venture across the river to buy a bottle of tequila and see "Old Mexico."

Not until the United States treats Mexico like the brother it has always been will the International Park become a reality. But time is running out, as it always seems to do, in the canyons of Big Bend.

Las Barrancas
del Cobre
Chihuahua

For centuries the great Sierra Madre Occidental has been one of
the major terra incognitae of our continent; a mysterious region
of immense gorges and pine covered mesas, of strange tribes,
lost mines, and romantic stories.

THOMAS B. HINTON
in Carl Lumholtz's *Unknown Mexico* (1902)

In the two decades since those words were penned for the new edition of
Carl Lumholtz's 1902 classic, that description still fits. If anything, the
legendary Sierra Madre, and its gaping, flood-strewn barrancas, are even
wilder than they were when the Norwegian explorer first crossed the hard,
myth-shrouded "Mother Mountains" with local Indian guides in 1890.

Described by geographers as an extension of the Rocky Mountains, the
Sierra Madre Occidental suddenly erupts below the southern borders of
Arizona and New Mexico and slashes its way southward across the
Mexican states of Sonora, Chihuahua, Sinaloa, Durango, and Nayarit in
one continuous 750-mile push; where this cloud-hugging massif forms the
Continental Divide and the natural border between the frontier states of
Sonora and Chihuahua lie the mysterious Las Barrancas del Cobre.

Frequently touted as being larger and deeper than the Grand Canyon,
Las Barrancas del Cobre actually comprise a Brobdingnagian region of

A Tarahumara woman waits for the Semana Santa, *Holy Week, celebration to begin in the Sierra Madre, Chihuahua.*

mile-deep canyons, forested sierras, and a cataclysmic mixture of lava, tufa, and ancient granites formed during the Tertiary period. Comprised of five distinct physiographic provinces, the 9,000-foot-high crest of the Sierra Madre del Norte lies within the Mexican Cordilleran Plateau province; and its dramatic western escarpment is drained by ancient rivers named after native peoples, like the Río Yaqui and Río Mayo, and others like the Río Fuerte and Río Sinaloa, which, like the Yaqui and Mayo rivers, also find their way to the Sea of Cortez. In so doing, these Sierran rivers—carrying the turbulent runoff of awesome tributaries like the Río Verde, Río Batopilas, and Río Urique, among others—have carved the Sierra Madre del Norte's greatest barrancas and created one of the most spectacular and tortured canyon regions on earth. Known as the Canyon Country province, it's gouged by 6,000-foot-deep trenches such as the Barranca de Urique, Barranca de Sinforosa, and Barranca de Batopilas, among others. East of the Sierra Madre del Norte's plateau-like crest, however, the gentler east slope of the Continental Divide is drained principally by the Río Conchos, which tumbles through the alluring Foothills, Plains, and Basin and Range provinces before finally emptying into the Río Bravo del Norte (known north of the border as the Río Grande).

But whether the Sierra Madre del Norte's magnificent rivers have created any barrancas deeper and larger than the Grand Canyon, as many still claim, is debatable. What's more striking is that the canyons of the Colorado River pale in comparison to the number of indigenous people still living in a 25,000-square-mile area of Las Barrancas del Cobre, much the way archaeologists tell us the Tarahumara first inhabited southwestern Chihuahua some 2,000 years ago. Totaling some 50,000 people today, the Uto-Aztecan-speaking Tarahumara may have been descendants of Arizona's Apache; at least, Padre Luis Velarde linked both tribes by name circa 1716 when he wrote, "On the opposite side and on this bank of the Jila [Gila River] also to the northeast live the Apache, the Pimas call them Tarasoma. . . ."

Whether the Tarahumara and the Tarasoma are actually blood kin, however, is not known. What *is* certain is that, next to the 165,000-strong Navajo nation, the Tarahumara are the second largest tribe in North America and they have long since called themselves *rarámuri*, "footrunners." For centuries, their fleet-footed prowess has enabled them to run a deer to death, or cover 200 miles at a trot before a second sun had set. In his ethnographic account of these storied runners, Jacob Fried grouped the *rarámuri* into three subcultural bands—a hardy people who frequently interchanged canyon and mountain environments as necessity and season

OVERLEAF: *A procession of Tarahumara women join the* Semana Santa *celebration, while other tribal members view them from the cliffs high above at Sierra Madre.*

Finalio Perez wears the traditional muslin clothes of the lowland Tarahumara in the Barranca del Batopilas, Chihuahua.

demanded. The *pagótame,* "mountain dwellers," comprise the majority of Tarahumaras and inhabit the Cordilleran Plateau province, a coniferous region covered with mesmerizing stands of Chihuahuan pine, Arizona pine, and Douglas-fir. Here, the *pagótame* traditionally live in caves and stone dwellings amongst lush mountain meadows and grassy basins where they grow corn, beans, and squash, graze cattle and goats, and hunt many animals on foot. At one time their larders also included deer they cornered, stabbed, and bludgeoned after a day- or two-day-long chase. The *poblanos,*

Two Tarahumara girls, Las Barrancas del Cobre, Mexico.

Swiss hiker Andri Rauch (left) helps British friend Shaun Cannon negotiate a leap during an exploratory trek down the flood-swollen Barranca del Cusárare.

"canyon dwellers," on the other hand, only make up an estimated 10–12 percent of the *rarámuri;* perhaps because they've traditionally scratched out a living in the more rugged and austere Canyon Country province—a surreal, Carlos Casteñedan world where the subtropics and Sonoran desert meet in a hallucinogenic mix of towering cacti, sweet mangos, tenacious century plants, and native palm trees. Here the *poblanos* live in caves among wild parrots and trogons, jaguar, and javelina, hunting and gathering like their highland cousins, tilling small earthen plots of maize for *pinole* (corn

*Modern Tarahumara pictograph in
Barranca del Cusárare, Chihuahua.*

gruel) and ritual *tesqüinada* (drinking ceremony) sometimes fishing swollen creeks and streams for bass, catfish, and squawfish. The *simaróne,* "wild people," however, dance in the shadows between the *poblanos'* and *pagótames'* worlds because they were the only band of *rarámuri* that successfully resisted conversion to Catholicism introduced throughout the Sierra Madre del Norte during the seventeenth century.

Like native peoples everywhere, much of the *rarámuri's* early history among Europeans was also written in blood, pestilence, and famine, and their cultural veins were first severed not long after Juan Fonte, a Jesuit, met with eight hundred fresh souls in 1607. Those who resisted conversion, the *gentile* (heathen) as well as the majority of *rarámuri* who eventually "saw the light" in some fashion or the other, nevertheless faced a relentless tide of *čabóči,* "non-Indians," that included Jesuit and Franciscan missionaries, cattlemen, miners, soldiers, *villistas,* and Apaches—an onslaught that forced many traditional Tarahumara to flee their ancestral homes in the pastoral lands on the east slope of the Sierra Madre del Norte and retreat deeper and deeper into the lofty sierras and rugged western barrancas.

One of the first *čabóčis* to meet these modern cave-dwellers on their terms, however, was a *blanco,* a white man named Carl Lumholtz. Between 1890 and 1898, Lumholtz criss-crossed the Sierra Madre Occidental on horseback with Tarahumara guides and came to know their people better than any ethnographer living then—or since. The highly acclaimed *Unknown Mexico* remains an enthralling record of Lumholtz's life with, and exploration among, the Tarahumara. Of Chihuahua's deepest canyon, the Barranca de Urique, Lumholtz wrote, "Even the intrepid Jesuit Missionaries at first gave up the idea of descending into it, and the Indians told them that only the birds knew how deep it was. The traveller as he stands at the edge of such gaps wonders whether it is possible to get across them."

U.S. entrepreneur Albert Kinsey Owen didn't wonder too long; first commissioned in 1849, Owen marshaled forces in 1872 to build the "World's Most Scenic Railroad" across the vast haunt of the Tarahumara, a region once roamed by the now extinct *cíbolo,* or bison, Chihuahuan grizzly, and Merriam elk. But the engineering marvel wasn't completed until November 21, 1921. That's when the 565-mile-long Ferrocarril Chihuahua al Pacífico finally linked Ojinaga, Chihuahua, with Los Mochis, Sinaloa, on the Sea of Cortez. To do so, however, the imposing 8,000-foot-high western escarpment of the Sierra Madre del Norte had to be breached, and seventy-two tunnels and twenty-seven bridges were initially needed before the linchpin 122 miles of track was finally hammered down a century after construction had first begun.

No other inroad has so affected the modern Tarahumara's traditional way of life. Even before construction was completed, glassy-eyed men in search of the lost Treasure of the Sierra Madre were feverishly dynamiting gloryholes into once-inaccessible sierras and barrancas; virgin stands of western yellow pine and Douglas-fir were leveled for commercial timber; and modern influences groped further and further into the timid *rarámuri*'s last stronghold, bringing change to a resolute people who'd desperately clung to their traditional life and spirit ways as no other North American tribe had been able to do.

In *Mexico: A Higher Vision* (1990, p. 19), Carlos Fuentes wrote, "This great Mexican abyss bears witness to the two extremes of creation: birth

Tarahumara women and children arrive en masse for Easter week's Semana Santa *celebration.*

and death." And when it's all said and done, the railroad will either be the cultural death knell for the Tarahumara, forcing them to flee to the desperate *poblaciones indigentes,* (indigent populations) of cities such as Chihuahua and Juárez, or to adopt the westernized *mestizo* ways of their neighbors. However, if they were right about the *čabóči,* "that God destroys the souls of *čabóči* who have been particularly mean during their lives in a fire he keeps for this purpose," the *rarámuri* might somehow remain culturally intact, in spite of the estimated 200,000 *čabóči*s of largely *mestizo* (Mexican and Indian) and Mexican descent now inhabiting and repopulating the *rarámuri*'s dwindling ancestral lands. With the North American Free Trade Agreement all but a done deal, though, their future does not look promising; chomping at the bit is a multinational array of business titans, pitchmen, handlers, and employees threatening to displace Tarahumara from their existing lands if the Mexican government ratifies its national constitution to eliminate the once protective *ejidos* (communal land system), in favor of exploiting the region's treasure trove of natural resources.

Schoolgirls, Creel, Chihuahua.

And that's not the only threat.

To truly understand the Tarahumara's precarious cultural dilemma, you need only look back on the United States' own frontier days; because traveling through the Barrancas del Cobre region today doesn't seem much different from what historians and Native American elders have told us traveling was like in the Wild West. If you take the "Most Spectacular Train Ride in the World" from Los Mochis, Sinaloa, to Creel, Chihuahua, you will thunder across the bioregional and cultural spectrum of the *poblano*, *pagótame*, and *simaróne* ancestral lands: from the dense thorn forests smothering the *bajadas* and western foothills of the Sierra Madre del Norte, through the dark subtropical barranca of the Río Septentrión to your first peek at the Barranca del Cobre from the yawning vista of El Divisadero. Here you will have your first look at "Mexico's Grand Canyon," home to twenty different species of oak and two prolific plants Americans can't get enough of: *la mota* (marijuana) and *la goma* (black tar heroin). Both drugs are changing the face of a native people and the barrancas they inhabit.

A Tiger lily blooms in the Barranca del Cusárare, Chihuahua.

Sixty klicks up the line from El Divisadero is Creel, pronounced *crill*, nestled in the forested highlands of the Continental Divide. Founded in 1907, Creel is the Dodge City of Chihuahua. One walk around the plaza during peak harvest—and tourist—season says it all. Traditionally dressed Tarahumara eat chips and drink sodas; while *narcotraficantes* work on the details of getting their product out of the sierra from the seats of heavily chromed 4WD American pickups. State and Federal Judicial Police, packing .45 automatics with the hammers already cocked, play cat and mouse with the *narcos*. DEA agents, and who really knows what other mix of U.S. operatives working under deep cover, try to mesh with local hombres, but the piece nestled in their lizard-skin boots, or in the small of their backs, usually gives them away in the flick of an eye. Bell Rangers, flown by Mexico City-based military pilots, launch from their helipads on the outskirts of town to spray uproquat on lush stands of *mota* and *amapola* (opium poppies); while schoolchildren stare out the windows every time a *helicóptero* returns from its mission in the sierra. Their parents work mostly

Two Tarahumara men dressed for the Semana Santa *celebration.*

A Tarahumara woman grinds corn for making tesqüino, *a fermented corn beer,
ritually used in sacred Tarahumara celebrations during* Semana Santa.

as woodcutters, sawmill workers, missionaries, *evangelistas*, miners, ranchers, Americans on the lam, and local merchants who cater to the burgeoning tourist industry. Everybody knows what's going on, except the foreign
tourists; during the summertime, they're mostly trilingual Europeans
speaking fluent Spanish heavily accented with German, Swiss, Italian,
Spanish, and British. All they want to do is see the Tarahumara in
storybook form. They don't know about the Tarahumara being conscripted
at gunpoint to harvest the dope to sooth the norteamericanos. And the
Tarahumara aren't saying—they know they'll get hit from both sides.
According to one report, "Tarahumaras talk about a time the soldiers
mistook a group of Tarahumaras for drug traffickers. The Indians were
machine-gunned from a flying helicopter. [But] Chihuahua authorities
said they could not recall any such incident" (Ruben Hernandez, "Backs to
the Wall," *Tucson Citizen,* January 31, 1989, p. 1-A).

OVERLEAF: *Surrounded by pine trees and fertile fields of maize, corn, a Tarahumara
man repairs the roof of his house in the highlands of the Sierra Madre, Chihuahua.*

*A young passenger gazes out the train window of the Ferrocarril Chihuahua al Pacífico
during a brief stop in Creel, Chihuahua.*

The story is probably true. There's no reason not to believe the
Tarahumara. Look at the guns in the plaza. This is the Sierra Madre, my
friend. Better yet, take the tour from Creel to Cascada de Basaseachic; it
will make a believer out of you. Located twelve hours of hard driving south
of Tucson, Arizona, on the new trans-Sierra, President Salinas de Gortori
Highway, Cascada de Basaseachic is the highest waterfall in Mexico. It
forms the headwaters for the Río Mayo, but this isn't like any waterfall

Tarahumara, mestizo, *and Mexican boys gaze in wonder at the television set used by a Ciudad Juarez, Chihuahua, television crew to videotape the* Semana Santa *celebration in a village of the Sierra Madre.*

you've ever seen before. Plunging 867 feet in a single, stomach-turning drop, you can look straight down this thundering cataract as it roars down Cañón del Candamena. And as you're lulled closer to the edge, you can't help but wonder if this is the vision the *rarámuri* have of the Mother Mountains without *čabóči.* Only, they aren't saying. But if a timid Tarahumara ever looks you dead in the eye after three days and nights of dancing the *dutúburi* (rain dance) during *semana santa,* Holy Week, you'll know it's the hallowed ground of places like Basaseachic that still course through their veins. And like its eternal waters hunting their way to the sea, through the now-abandoned barrancas of the *rarámuri,* you'll pray the Tarahumara find a way to keep what the rest of us have already lost.

Lost Canyons of the
Ancient Ones
Colorado, New Mexico, and Utah

The spirits lingering in the abandoned hunting camps . . . have
an eloquence of their own—if we will only listen. How does one
recapture the excitement permeating a hunting band that has just
succeeded in bringing a mammoth to earth . . . with stone-
tipped spears and the cunning and bravery of the hunters?

EMIL W. HAURY
Those Who Came Before (1983)

You don't recapture the excitement of the hunt, not 10,000 years after the
beast has fallen, the bones have been picked clean, the hide has withered,
and the remains have turned to dust. But you can always listen to the
stones, the ancient dwellings that still cling to them, the voices that once
rang out from these canyon homes. They are everywhere, and they are
nowhere, but they have a story to tell—if we will only listen. Tucked into
the secret labyrinths and lost canyons of the Anasazi, "ancient people,"
they are the only evidence modern Southwest peoples now have that the
region really wasn't what Spanish padres concluded long ago was *despoblado,* "uninhabited." It has always been inhabited—at least as far back
as the great mammoth hunters; that's when man first realized one frightening option stood between him and extinction: bring down a rampaging
mastodon with sharp spears flung from *atlatls.*

But that was the very beginning, when these first Southwesterners spoke a language lost to myth, when their spiritual visions were woven into their eloquent baskets and painted on their fragile *ollas*. To glean anything more, you need to follow the ancient corn path north out of Tehuácan Valley of central Mexico to see that it also led to the lost canyons of the Anasazi. To reach the Anasazi over two thousand miles north, corn kernels were passed from one ancient culture to the next. First along this 5,000-year-old trail lived those whom some have called the Chichimecas; these ancient people inhabited the high mountains and deep barrancas of the Sierra Madre Occidental. And, unlike the abandoned canyons of the Anasazi, the caves and stone dwellings of the Chichimeca ancestral ground are still inhabited by their descendants—the Tarahumara.

Farther north, beyond the flaming deserts of the Hakataya and the Hohokam, and the piney highlands of the Mogollon peoples, maize finally reached the Anasazi. The story is written in the ancient dwellings, the cliff palaces of Mesa Verde, the stone towers of Hovenweap, the clustered adobes of Bandelier, and the caves, the rock shelters, the crude windbreaks of those beyond. They are the remains of a wise people who once lived, breathed, and stalked the high ground and deep canyons of the Colorado Plateau for at least 1,400 years. Where we see the sun, moon, and stars, they had visions—and they left the trails of the cosmos etched in stone. Like us, they also had families and societies, but they were bound by spirit, ritual, and legend, not by "advanced" governments and religions. And while industrialized societies live where they want and take what they want, with little regard for the planet and the Third World peoples also struggling to survive on it, the Anasazi lived closer to the earth than modern humanity can ever imagine. They had an encyclopedic knowledge of plants and knew both their practical and medicinal uses; and they knew how to hunt the great bison, the deer, and the antelope—and when to retreat to their canyon dwellings to plant maize, beans, and squash.

Archaeologists, ethnographers, anthropologists, and all manner of non-Indian scientists have been studying these ancient matters since they first "discovered" the abandoned dwellings of the lost canyons. But the ancestral knowledge still lies with the descendants of the great Anasazi culture; it's in the heartbeat of native elders, the eyes of surviving medicine men. They are the spiritual and blood kin to an ancient people who occupied the Four Corners region until about A.D. 1300; then they vanished. And non-Indians do not know why, because they never heard the native voices.

Yet non-Indians are determined to find out; they are drawn to the great dwellings like lemmings. Why? What do they hope to see? And why are so many of these monuments devoid of the native peoples who could surely shed some insight into these matters if, indeed, non-Indians belong

Square Tower House is the tallest cliff dwelling in Mesa Verde National Park, Colorado.

A young visitor exploring the Long House ruins in Bandelier National Monument, New Mexico.

there at all? Non-Indians think they have a better way; it's called *science.* So they are left alone, to stare blankly into a past to which they have no cultural link: some hoping for answers, others merely looking for entertainment.

The late anthropologist Clyde Kluckhorn wrote that anthropology was a "mirror for man." If that's true, it's a mirror in which modern man can't see his reflection. But it's there in the lost canyons of the Anasazi for everyone to see: beautiful, magnificent dwellings—some even say cosmic. But they were abandoned. Why? Drought? Did the population outstrip the natural carrying capacity of the canyons and mesas? Did hostile tribes drive them off? Was it disease? It could be all of these; it might only be one of them. But if the question can't be answered, the empty caves and the abandoned dwellings should at least be a clue for how modern people should now live on planet Earth with one another. Go see for yourself. A

The Anasazi carved shelters, such as Long House ruins, into the soft rock walls lining
El Río de los Frijoles, *Frijoles River, circa* A.D. 1100.

Cascada de los Frijoles, *Lower Frijoles Falls, in Cañón de los Frijoles. Bandelier National Monument, New Mexico.*

Moonrise over Hovenweap Castle, built circa A.D. 1200, Hovenweap National Monument, Utah.

Taos Pueblo man said, "We have lived upon this land from the days beyond history's records, far past any living memory, deep into the time of legend. The story of my people and the story of this place are one single story. No man can think of us without thinking of this place. We are always joined together." No one can now see the earth, the deep canyons that still cleave it, without thinking of the native people who lived there; they were always joined together.

Bibliography

CANYON LANDS OF THE HIGH PLATEAU: UTAH AND ARIZONA

Annerino, John, with photographs by the author. "Sacred Mountains of the Navajo." *Arizona Magazine*, May 9, 1982, pp. 5–8.

Begay, Scotty, and Richard F. Van Valkenburgh. "Sacred Places and Shrines of the Navajo. Part I: The Sacred Mountains." *Museum Notes, Museum of Northern Arizona* 11, no. 3 (September 1938).

Bolton, Herbert E., trans. and annot. *Pageant in the Wilderness: The Story of the Escalante Expedition to the Interior Basin, 1776.* Salt Lake City: Utah State Historical Society, 1950.

Callaway, Donald, Joel Janetski, and Omer C. Stewart. "Ute." *Handbook of North American Indians.* Vol. 11: *Great Basin.* Washington, DC: Smithsonian Institution, 1986.

Crampton, C. Gregory. *Standing Up Country: The Canyon Lands of Utah and Arizona.* New York: Knopf, and University of Utah Press, 1964.

Doolittle, Jerome, with photographs by Wolf Von Dem Bussche. *Canyons and Mesas.* The American Wilderness Series. New York: Time-Life Books, 1974.

Fradkin, Philip L., with photographs by the author. *A River No More: The Colorado River and the West.* New York: Knopf, 1981.

Gregory, Herbert E., *Geology of the Navajo Country: A Reconnaissance of Parts of Arizona, New Mexico, and Utah.* Professional Paper 93. Washington, DC: U.S. Government Printing Office, 1917.

Gregory, Herbert E., *The San Juan Country: A Geographic and Geologic Reconnaissance of Southeastern Utah.* Professional Paper 188. Washington, DC: U.S. Government Printing Office, 1938.

Jenkinson, Michael, with photographs by Karl Kernberger. *Land of Clear Light: Wild Regions of the American Southwest—How to Reach Them and What You Will Find There.* New York: Dutton, 1977.

Kelly, Charles, "The Mysterious 'D. Julien.'" *Utah Historical Quarterly* 6, no. 3 (July 1933): 83–88.

McKnight, Edwin T. *Geology of Area Between Green and Colorado Rivers, Grand and San Juan Counties, Utah.* Bulletin 908. Washington, DC: U.S. Government Printing Office, 1940.

Powell, John W. *The Cañons of the Colorado.* Golden, Colorado: Outbooks, 1981. First published in Scribner's Monthly magazine in 1875: 9: 293–310, 394–409, 523–37.

Trimble, Stephen. *The Bright Edge: A Guide to the National Parks of the Colorado Plateau.* Flagstaff: Museum of Northern Arizona Press, 1979.

Van Valkenburgh, Richard F. *Diné Bikéyah* ("The Navajo's Country"). Window Rock, AZ: U.S. Department of the Interior, Office of Indian Affairs Navajo Service, 1941.

Van Valkenburgh, Richard F. *Navajo Indians III: Navajo Sacred Places.* New York: Garland, 1974.

Watson, Edith L. *Navajo Sacred Places.* Window Rock, AZ: Navajo Tribal Museum, 1964.

BLACK CANYON OF THE GUNNISON: COLORADO

Atwood, Wallace W., and Wallace W. Atwood Jr. "Working Hypothesis for the Physiographic History of the Rocky Mountain Region." *Bulletin of The Geological Society of America* 49 (June 1, 1939): 957–980.

Baggs, Mae Lacy. *Colorado: The Queen Jewel of the Rockies.* Boston: Page, 1918.

Beidleman, Richard G. "The Gunnison River Diversion Project." *Colorado Magazine* (State Historical Society of Colorado) 36, no. 3 (July 1959): 187–285.

Beidleman, Richard G. "The Black Canyon of the Gunnison National Monument." *Colorado Magazine* (State Historical Society of Colorado) 40, no. 3 (July 1963): 161–178.

Callaway, Donald, Joel Janetski, and Omer C. Stewart. "Ute." *Handbook of North American Indians.* Vol. 11: *Great Basin.* Washington, DC: Smithsonian Institution, 1986.

Hansen, Wallace R. *The Black Canyon of the Gunnison: In Depth.* Tucson: Southwest Parks & Monuments Association, 1987. First published as U.S. Geological Survey Bulletin 1191, 1874.

Peale, A. C. "Report of A. C. Peale, M.D.: Geologist of Middle Division, 1874." In F. V. Hayden, *Eighth Annual Report of the United States Geological and Geographical Survey of the Territories, Embracing Colorado and Parts of Adjacent Territories.* Washington, DC: U.S. Government Printing Office, 1876.

Schiel, Jacob H. Trans. (from German) and ed., Thomas N. Bonner. *Journey Through the Rocky Mountains and the Humboldt Mountains to the Pacific Ocean.* Norman: University of Oklahoma Press, 1959. Originally published in 1859 as *Reise durch die Felsengebirge und die Humboldtgebirge nach dem Stillen Ocean*, by Schaffhausen, Brodtmann'schen Buchhandlung.

Warner, Mark T. "Through the Canyon." *Colorado Magazine* (State Historical Society of Colorado) 40, no. 3 (July 1963): 179–182.

THE CANYONS OF ZION: UTAH

Annerino, John. Unpublished "Field Notes (with Maps): Explorations in Zion National Park and Canyonlands National Park, Utah; and the Black Canyon of the Gunnison National Monument, Colorado," 1985, and 1991–1992.

Dellenbaugh, Frederick S. *A Canyon Voyage: The Narrative of the Second Powell Expedition.* Tucson: University of Arizona Press, 1984. First published by Putnam in 1908.

Gregory, Herbert E., and Norman C. Williams. *Zion National Monument, Utah.* Springerville, UT: Art City Publishing, 1949. First published in the *Bulletin of the Geological Society of America* 58 (March 1947).

James, George Wharton. *Utah: The Land of Blossoming Valleys.* Boston: Page Company, 1922.

Kelly, Isabel T., and Catherine S. Fowler. "Southern Paiute." *Handbook of North American Indians*. Vol. 11: *Great Basin*. Washington, DC: Smithsonian Institution, 1986.

Palmer, W. R. "Indian Names in Utah Geography." *Utah Historical Quarterly* 1, no. 1. (January 1928): 5, 12–13, 16–17.

Palmer W. R. "Utah Indians Past and Present; An Etymological and Historical Study of Tribes and Tribal Names from Original Sources by Wm. R. Palmer, Cedar City, Utah." *Utah Historical Quarterly* 1, no. 2 (April 1928).

Powell, Maj. John Wesley. *Exploration of the Colorado River of the West, and Its Tributaries. Explored in 1869, 1870, 1871, and 1872.* Washington, DC: U.S. Government Printing Office, 1875.

Powell, Maj. John Wesley. *An Overland Trip to the Grand Cañon*. Palmer Lake, CO: Filter Press, 1974. First published in 1875 by Scribner's Monthly magazine: 10: 659–78.

Woodbury, Angus M. "A History of Southern Utah and Its National Parks." *Utah State Historical Society* 12, nos. 3–4 (July–October 1944). Revised and reprinted in 1950.

Woodbury, Angus M. "The Route of Jedediah S. Smith in 1826, from the Great Salt Lake to the Colorado River." *Utah Historical Quarterly* 4, no. 2 (April 1931).

CANYONS OF PARIA: UTAH AND ARIZONA

Annerino, John. *Adventuring in Arizona*. San Francisco: Sierra Club Books, 1991.

Annerino, John with photographs by Christine Keith. "Untamed and Tantalizing: Running Paria Canyon." *Running*, 9, no. 1 (February 1983): 42–46.

Brookes, Juanita, and Robert Glass Cleland. *A Mormon Chronicle: The Diaries of John D. Lee, 1848–1876*. San Marino, CA: Huntington Library, 1955.

Crampton, Gregory C. *Standing Up Country: The Canyon Lands of Utah and Arizona*. New York: Knopf, 1964.

Gregory, Herbert E., and Robert C. Moore. *The Kaiparowits Region: A Geographic and Geologic Reconnaissance of Parts of Utah and Arizona*. Washington, DC: U.S. Government Printing Office, 1931.

Kelly, Isabel T. "Southern Paiute Ethnography." University of Utah, Anthropological Papers, No. 69, May 1964 Glen Canyon Series Number 21.

Kelly, Isabel T., and Catherine S. Fowler. "Southern Paiute." *Handbook of North American Indians*. Vol. 11: *Great Basin*. Washington, DC: Smithsonian Institution, 1986.

Kelsey, Michael R. *Hiking and Exploring the Paria River*. Provo, UT: Kelsey, 1987.

McLuhan, T. C., compiler. *Touch the Earth: A Self-Portrait of Indian Existence*. New York: Promontory Press, 1971.

Reilly, P. T. "Historic Utilization of Paria River." *Utah Historical Quarterly* 45, no. 2 (Spring 1977): 188–201.

Rusho, W. L., and Gregory C. Crampton. *Desert River Crossing: Historic Lee's Ferry on the Colorado River*. Salt Lake City: Peregrine Smith, 1975.

T. V. B. "An Episode of Military Explorations and Surveys." *United Service: A Monthly Review of Military and Naval Affairs* 5 (1881): 468–472.

CANYONS OF REDROCK–SACRED MOUNTAIN: ARIZONA

Annerino, John. *Adventuring in Arizona*. San Francisco: Sierra Club Books, 1991.

Bartlett, Katherine. "Notes upon the Routes of Espejo and Farfan to the Mines in the Sixteenth Century." *New Mexico Historical Review* 108 (1942): 21–36.

Bolton, Herbert Eugene, ed. *Spanish Exploration in the Southwest: 1542–1706*. New York: Scribner's, 1916.

Basso, Keith H. "Western Apache." *Handbook of North American Indians:* Vol. 10: *Southwest*. Washington, DC: Smithsonian Institution, 1983.

Colton, Harold S. "A Brief Survey of the Early Expeditions into Northern Arizona." *Museum Notes of the Museum of Northern Arizona* 2, no. 9 (March 1, 1930): 1–4.

Farish, Thomas Edwin. *History of Arizona*. Vol. 5. Phoenix: © Thomas Edwin Farish, 1918.

Hammond, George Peter, and Agapito Rey. *Expedition into New Mexico Made by Antonio de Espejo, 1582–1583: As Revealed in the Journal of Diego Pérez de Luxán, a Member of the Party*. Los Angeles: Quivira Society, 1929.

Huff, Lyman C., Elmer Santos, and R. G. Raabe. "Mineral Resources of the Sycamore Canyon Primitive Area Arizona." Geological Survey Bulletin 1230-F. Washington DC: U.S. Government Printing Office, 1966.

James, George Wharton, *Arizona: The Wonderland*. Boston: Page Company, 1917.

Khera, Sigrid, and Patricia S. Mariella. "Yavapai." *Handbook of North American Indians*. Vol. 10: *Southwest*. Washington, DC: Smithsonian Institution, 1983.

Lowe, Charles H. *Arizona's Natural Environment*. Tucson: University of Arizona Press, 1964.

Thrapp, Dan L. *Al Sieber: Chief of Scouts*. Norman: University of Oklahoma Press, 1964.

Wells, Edmund. *Argonaut Tales: Stories of the Gold Seekers and the Indian Scouts of Early Arizona*. New York: Hitchcock, Grafton Press, 1927.

GRAND CANYON OF THE COLORADO RIVER: ARIZONA

Annerino, John. "National Park Cops: Keeping the Peace in Our National Parks." *Police* 12, no. 11 (November 1988): 36–43.

Annerino, John. *High Risk Photography: The Adventure Behind the Image*. Helena: American & World Geographic Publishing, 1991.

Annerino, John, with photographs by Christine Keith. *Running Wild: Through the Grand Canyon, on the Ancient Path*. Tucson: Harbinger House, 1992.

Annerino, John. *Hiking the Grand Canyon*. Rev. and expanded. San Francisco: Sierra Club Books, 1993. First published in 1986.

Brew, J. O. "Hopi Prehistory and History to 1850." *Handbook of North American Indians*. Vol. 9: *Southwest*. Washington, DC: Smithsonian Institution, 1979.

Brown, David E., ed. *The Wolf in the Southwest: The Making of an Endangered Species*. Tucson: University of Arizona Press, 1983.

Brugge, David M. "Navajo Prehistory and History to 1850." *Handbook of North American Indians*. Vol. 10: *Southwest*. Washington, DC: Smithsonian Institution, 1983.

Coues, Elliott, ed. and trans. *On the Trail of the Spanish Pioneer: The Diary and Itinerary of Francisco Garces, 1775–1776*. New York: F. P. Harper, 1900.

Dawson, Thomas F. *First Through the Grand Canyon*. U.S. Senate Resolution No. 79, June 4, 1917.

Dellenbaugh, Frederick S. *Romance of the Colorado River*. New York: Putnam's, 1906.

Dellenbaugh, Frederick S. *A Canyon Voyage: The Narrative of the Second Powell Expedition Down the Green-Colorado River from Wyoming, and the Explorations on Lands, in the Years 1871 and 1872*. New Haven, CT: Yale University Press, 1926.

Dutton, Clarence E. *Tertiary History of the Grand Canyon District, With Atlas.* Washington, DC: U.S. Government Printing Office, 1882.

Euler, Robert C., and A. Trinkle Jones. *A Sketch of Grand Canyon Prehistory.* Grand Canyon, AZ: Grand Canyon Natural History Association, 1979.

Hughes, J. Donald. *The Story of Man at the Grand Canyon.* Grand Canyon, AZ: Grand Canyon Natural History Association, 1967.

Ives, Joseph Christmas. *Report upon the Colorado River of the West: Explored in 1857 and 1858.* Washington, DC: U.S. Government Printing Office, 1861.

James, George Wharton. *In and Around the Grand Canyon.* Boston: Little, Brown, 1900.

James, George Wharton. *The Grand Canyon of Arizona: How to See It.* Boston: Little, Brown, 1910.

Kelly, Isabel T., and Catherine S. Fowler. "Southern Paiute." *Handbook of North American Indians.* Vol. 10: *Great Basin.* Washington, DC: Smithsonian Institution, 1986.

Lavender, David Sievert. *River Runners of the Grand Canyon.* Grand Canyon, AZ: Grand Canyon Natural History Association, 1985.

Lingenfelter, Richard E. *First Through the Grand Canyon.* Los Angeles: Glen Dawson, 1958.

McGuire, Thomas R. "Walapai." *Handbook of North American Indians.* Vol. 10: *Southwest.* Washington, DC: Smithsonian Institution, 1983.

Peattie, Roderick, ed. *The Inverted Mountains: Canyons of the West.* New York: Vanguard Press, 1948.

Powell, Maj. John Wesley. *Exploration of the Colorado River of the West, and Its Tributaries. Explored in 1869, 1870, 1871, and 1872.* Washington DC: U.S. Government Printing Office, 1875.

Reisner, Marc. *Cadillac Desert: The American West and Its Disappearing Water.* New York: Viking, 1986.

Schwartz, Douglas W. "Havasupai." *Handbook of North American Indians.* Vol. 10: *Southwest.* Washington, DC: Smithsonian Institution, 1983.

Smith, Melvin T. *The Colorado River: Its History in the Lower Canyons Area.* Provo, UT: Brigham Young University, 1972.

Spanner, Earle E., ed. *Bibliography of the Grand Canyon and the Lower Colorado River: 1540–1980.* Monograph No. 2. Grand Canyon, AZ: Grand Canyon Natural History Association, 1981.

Whitney, Stephen, with illustrations by the author. *A Field Guide to the Grand Canyon.* New York: William Morrow-Quill, 1982.

CAÑÓN DEL DIABLO: BAJA CALIFORNIA NORTE

Beal, Carl H. "Reconnaissance of the Geology and Oil Possibilities of Baja California, Mexico." Geological Society of America, Memoir 31, December 1, 1948.

Bolton, Herbert Eugene, ed. *Kino's Historical Memoir of Pimería Alta.* Berkeley and Los Angeles: University of California Press, 1948.

Burrus, Ernest J., trans. and ed. *Wenceslaus Linck's Diary of His 1766 Expedition to Northern Baja.* Los Angeles: Dawson's Book Shop, 1966.

Cudahy, John. *Mañanaland: Adventuring with Camera and Rifle Through California in Mexico.* New York: Duffield, 1928.

Ekholm, Gordon F., and Gordon R. Willey, vol. eds. *Handbook of Middle American Indians.* Vol. 4: *Archaeological Frontiers and External Connections.* Austin: University of Texas Press, 1966.

Johnson, William Weber. *Baja California.* The American Wilderness Series. New York: Time-Life Books, 1972.

McQuown, Norman A., vol. ed. *Handbook of Middle American Indians* Vol. 5: *Linguistics.* Austin: University of Texas Press, 1967.

Nelson, Edward W. "Lower California and Its Natural Resources." Vol. 106: *First Memoir.* National Academy of Sciences. 1966.

Robinson, Bestor. "The Ascent of El Picacho del Diablo." *Sierra Club Bulletin* 108, no. 1 (February 1933): 56–63.

Robinson, John W. *Camping and Climbing in Baja.* Glendale, CA: La Siesta Press, 1983.

Vogt, Evon Z., vol. ed. *Handbook of North American Indians.* Vol. 8: *Ethnology,* Part 2. Austin: University of Texas Press, 1969.

Woodford, A. O., and T. F. Harriss. "Geological Reconnaissance Across Sierra San Pedro Martir, Baja California." *Bulletin of the Geological Society of America* 49 (September 1, 1938).

CAÑONES DE LA ISLA TIBURÓN: SONORA

Annerino, John. "People of the Dune." *Northern Arizona Life* May 1985, pp. 24–32.

Bowden, Charles. *Desierto: Memories of the Future.* New York and London: W. W. Norton & Co., 1991.

Bowen, Thomas. *Seri Prehistory: The Archaeology of the Central Gulf Coast of Sonora, Mexico.* Tucson: University of Arizona Press, 1976.

Bowen, Thomas. "Seri." *Handbook of North American Indians.* Vol. 10: *Southwest.* Washington, DC: Smithsonian Institution, 1983.

Carmony, Neil B., and David E. Brown, eds. *Mexican Game Trails: Americans Afield in Old Mexico, 1866–1940.* Norman: University of Oklahoma Press, 1991.

Carmony, Neil B., and David E. Brown, eds. *Tough Times in Rough Places: Personal Narratives of Adventure, Death, and Survival on the Western Frontier.* Silver City, NM: High Lonesome Books, 1992.

Carmony, Neil B., and David E. Brown, eds. *Tales of Tiburon: An Anthology of Adventures in Seriland.* Phoenix: Southwest Natural History Association, 1983.

Felger, Richard S., and Mary Beck Moser. *People of the Desert and the Sea: Ethnobotany of the Seri Indians.* Tucson: University of Arizona Press, 1985.

Heyerdahl, Thor. *The Ra Expeditions.* London: Allen & Unwin, 1971.

Hills, Richard James. "An Ecological Interpretation of Prehistoric Seri Settlement Patterns in Sonora, Mexico." Unpublished master's thesis. Tempe: Arizona State University, 1973.

Ives, Ronald L. *Land of Lava, Ash, and Sand: The Pinacate Region of Northwestern Mexico.* Tucson: Arizona Historical Society, 1989.

Sheldon, Charles. *The Wilderness of Desert Bighorns and Seri Indians: The Southwestern Journals of Charles Sheldon.* Phoenix: Arizona Desert Bighorn Sheep Society, 1979.

THE CANYONS OF THE BIG BEND: TEXAS AND COAHUILA

Annerino, John. Unpublished "Field Notes (with maps): Explorations in the Sierra del Carmen, Coahuila; Barrancas del Cobre, Chihuahua; Isla Tiburón, Sonora; and Sierra San Pedro Mártir, Baja California Norte." 1972 and 1990–1992.

Calderwood, Michael, and Gabriel Breña, with aerial photographs by Michael Calderwood. *Mexico: A Higher Vision.* La Jolla, CA: Alti, 1990.

Emory, William H. *Report on the United States and Mexico Boundary Survey.* Vol. 1. Washington, DC: Cornelius Wendell, 1857.

Griffen, William B. "Southern Periphery: East." *Handbook of North American Indians.* Vol. 10: *Southwest.* Washington, DC: Smithsonian Institution, 1983.

Hall, Douglas Kent, with photographs by the author. *The Border: Life on the Line.* New York: Abbeville Press, 1988.

Horgan, Paul. *Great River: The Rio Grande in North American History.* Vols. 1 and 2. New York: Rinehart, 1954.

Jackson, Donald Dale, and Peter Wood, with photographs by Dan Budnick. *The Sierra Madre.* The American Wilderness Series. New York: Time-Life Books, 1975.

Kingston, Mike, ed. *Texas Almanac: 1990–91.* Dallas: Dallas Morning News, 1989.

Poppa, Terrence E. *Drug Lord: The Life and Death of a Mexican Kingpin.* New York: Pharos Books, 1990.

Tyler, Ronnie C. *The Big Bend: A History of the Last Texas Frontier.* Handbook 128. Washington, DC: U.S. Department of the Interior, 1975.

Wauer, Roland H. *A Naturalist's Mexico.* College Station: Texas A & M University Press, 1992.

Weisman, Alan, with photographs by Jay Dusard. *La Frontera: The United States Border with Mexico.* San Diego: Harcourt, Brace, Jovanovich, 1986.

Wuerthner, George. *Texas' Big Bend Country.* Helena, MT: American Geographic Publishing, 1989.

Zwinger, Ann Haymond. *The Mysterious Lands.* New York: Truman Talley Books/Plume, 1989.

LAS BARRANCAS DEL COBRE: CHIHUAHUA

Bennett, Wendell C., and Robert M. Zingg. *The Tarahumara: An Indian Tribe of Northern Mexico.* Glorieta, NM: 1976. First published in Chicago: University of Chicago, 1935.

Brand, Donald D. "The Historical Geography of Northwestern Chihuahua." Unpublished doctoral dissertation, University of California, 1933.

Brown, David E., and Neil B. Carmony. *Gila Monster: Facts and Folklore of America's Aztec Lizard.* Silver City, NM: High Lonesome Books, 1991.

Carmony, Neil B., ed. *Afield with Frank Dobie: Tales of Critters, Campfires, and the Hunting Trail.* Silver City, NM: High Lonesome Books, 1992.

Fayhee, M. John. *Mexico's Copper Canyon Country: A Hiking and Backpacking Guide to Tarahumara-land.* Evergreen, CO: Cordillera Press, 1989.

Fisher, Richard D. *National Parks of Northern Mexico.* Tucson: Sunracer Publications, 1990.

Fontana, Bernard L., with photographs by Helga Teiwes. *The Material World of the Tarahumara Indians.* Flagstaff: Arizona State Museum/University of Arizona, 1979a.

Fontana, Bernard L., with photographs by John P. Schaefer. *Tarahumara: Where Night is Day of the Moon.* Flagstaff: Northland Press, 1979b.

Fried, Jacob. "The Tarahumara." *Handbook of Middle American Indians.* Vol. 8: *Ethnology,* Part 2. Austin: University of Texas Press, 1969.

Hovey, Edmund Otis. "A Geological Reconnaissance in the Western Sierra Madre of the State of Chihuahua, Mexico." *Bulletin of the American Museum of Natural History* 203 (June 1907).

King, Robert E. "Geological Reconnaissance in Northern Sierra Madre Occidental of Mexico." New York: *Bulletin of the Geological Society of America* 50, no. 11 (November 1, 1939).

Llaguno, Rodrigo J. "Change in the Tarahumara Family: The Influence of the Railroad." Unpublished master's thesis, Louisiana State University, 1971.

Lumholtz, Carl. *Unknown Mexico: A Record of Five Years' Exploration Among Tribes of The Western Sierra Madre; in the Tierra Caliente of Tepic and Jalisco; and Among the Tarascos of Michoacan.* (With map.) New York: AMS Press, 1973. First published in New York: Scribner's, 1902.

Merrill, William L. "Tarahumara Social Organization, Political Organization, and Religion." *Handbook of North American Indians.* Vol. 10: *Southwest.* Washington, DC: Smithsonian Institution, 1983.

Merrill, William L., *Rarámuri Souls: Knowledge and Social Process in Northern Mexico.* Washington, DC: Smithsonian Institution Press, 1988.

Pennington, Campbell W. *The Tarahumar of Mexico: Their Environment and Material Culture.* Salt Lake City: University of Utah Press, 1963.

Pennington, Campbell W. "Tarahumara." *Handbook of North American Indians.* Vol. 10: *Southwest.* Washington, DC: Smithsonian Institution, 1983.

Running, John. *Honor Dance: Native American Photographs.* Reno: University of Nevada Press, 1985.

Shepherd, Grant. *The Silver Magnet: Fifty Years in a Mexican Silver Mine.* New York: Dutton, 1938.

Spicer, Edward H. "Northwest Mexico: Introduction." *Handbook of Middle American Indians.* Vol. 8: *Ethnology,* Part 2. Austin: University of Texas Press, 1969.

LOST CANYONS OF THE ANCIENT ONES: COLORADO, NEW MEXICO, AND UTAH

Cordell, Linda S. "Prehistory: Eastern Anasazi." Washington, DC: Smithsonian Institution, 1979.

Di Peso, Charles. "Prehistory: O'otam." Washington, DC: Smithsonian Institution, 1979.

Flog, Fred. "Prehistory: Western Anasazi." Washington, DC: Smithsonian Institution, 1979.

Gummerman, George J., and Emil W. Haury. "Prehistory: Hohokam." Washington, DC: Smithsonian Institution, 1979.

Jackson, W. H. "Ancient Ruins in Southwestern Colorado." In F. V. Hayden, *Eighth Annual Report of the United States Geological and Geographical Survey of the Territories, Embracing Colorado and Parts of Adjacent Territories.* Washington, DC: U.S. Government Printing Office, 1876.

Lister, Robert H., and Florence C. Lister, with photographs from the George A. Grant Collection and David Muench. *Those Who Came Before.* Tucson: Southwest Parks and Monuments Association and the University of Arizona Press, 1983.

Martin, Paul S. "Prehistory: Mogollon." Washington, DC: Smithsonian Institution, 1979.

Ortiz, Alfonso. "Introduction." *Handbook of North American Indians.* Vol. 9: *Southwest.* Washington, DC: Smithsonian Institution, 1979.

Schroeder, Albert H. "Prehistory: Hakataya." Washington, DC: Smithsonian Institution, 1979.

Woodbury, Richard B. "Prehistory: Introduction." *Handbook of North American Indians.* Vol. 9: *Southwest.* Washington, DC: Smithsonian Institution, 1979.

Zubrow, Ezra B. W., and Richard B. Woodbury. *Agricultural Beginnings, 2000 B.C.–A.D. 500.* Washington, DC: Smithsonian Institution, 1979.